Advanced Chemistry
Multiple Choice Tests

By the same author
Chemistry Data Book: SI Edition   co-author H. G. Wallace
Basic Chemistry Data Book
Titrimetric Analysis for A and S Levels
Questions and Problems in Inorganic Chemistry

Modern Chemistry Background Readers, edited by J. G. Stark

Chemical Periodicity   D. G. Cooper
Electrochemistry   P. D. Groves
Inorganic Complexes   D. Nicholls
The Shapes of Organic Molecules   N. G. Clark

Advanced Chemistry Multiple Choice Tests

J. G. Stark
Head of Chemistry Department, Glasgow Academy

John Murray   Albemarle Street   London

© J. G. Stark 1978

All rights reserved. No part of this publication may be reproduced, stored in a retrieval system, or transmitted, in any form or by any means, electronic, mechanical, photocopying, recording, or otherwise, without the prior permission of John Murray (Publishers) Ltd, 50 Albemarle Street, London W1X 4BD

Set on Linotron Filmsetter and printed in Great Britain by
J. W. Arrowsmith Ltd, Bristol

ISBN 0 7195 3383 X

Contents

To the teacher vi
To the student viii
Acknowledgements and For further reading ix
Specimen answer sheet x
Directions for answering the different types of item xi
Data xii

1. Atomic structure 1
2. Structure and bonding 8
3. The periodic table 13
4. States of matter 19
5. Energetics 25
6. Phase equilibria 32
7. Chemical equilibrium 1 40
8. Chemical equilibrium 2: redox equilibria and electrochemistry 47
9. Chemical equilibrium 3: acid–base equilibria 53
10. Rates of chemical reactions 60
11. s- and p-block elements 68
12. d-block elements 73
13. Organic chemistry 1: structure and bonding 80
14. Organic chemistry 2: chemistry of functional groups 87
15. Organic chemistry 3: large molecules 91
16. Revision test 98

Keys and facility indices (removable) 109

To the teacher

The use of objective tests in Advanced chemistry examinations has become more widespread in recent years and many examination boards have now built up a bank of items which is usually kept 'secure', though some boards are no longer maintaining the secrecy of their A-level multiple choice paper—for example, the University of London School Examinations Council which is responsible for the Nuffield Chemistry examinations.

An objective item may be defined as one for which there is only one predetermined correct answer. The commonest type is the multiple choice item consisting of a *stem* (an introductory question or an incomplete statement) and four or five[a] *responses* (alternative answers or completions), only one of which is correct (the *key*). The incorrect responses are called *distractors*. The multiple choice papers in this book are designed for the use of students following any modern A-level course, such as that of the Nuffield Foundation, the University of London or the Joint Matriculation Board. Each of the 15 test papers, classified according to subject area, consists of 20 items so that it can be completed (and possibly marked) in a single 40-minute class period or a similar homework period. In addition there is a 50-item revision test for which the suggested time allowance is $1\frac{1}{4}$ hours. A specimen answer sheet can be found on page x.

Four different types of item are used:

(a) multiple choice,
(b) multiple completion,
(c) classification and situation sets,
(d) assertion–reason.

These are more fully described on page xi. Only a limited number of assertion–reason items is included as this type of item is not as widely used in A-level examinations. A simplified version of Bloom's classification of educational objectives is used:

---

[a] Five in this book

To the teacher

1. recall (or knowledge),
2. comprehension (or understanding),
3. higher abilities (e.g. application, analysis).

The ability classification may vary from one set of students to another according to their previous learning experience and there may also be some overlap—the classification is hierarchical in nature. The following overall percentage of items measuring each ability is therefore only approximate:

1. 30%
2. 45%
3. 25%

All the items were pretested by students near the end of their A-level course. The number of students taking part in the pretesting of each paper was between 50 and 75. Item analysis was carried out to determine how easy or difficult the sample of students found each item and the ability of each item to discriminate between the able and less able students. The *facility index F* (the fraction of candidates who answer the item correctly) measures the former; the *discrimination index D* (defined as $F_h - F_l$ where F_h denotes the facility index for the highest 33% of candidates based on their total score and F_l the facility index for the lowest 33%) measures the latter. The possible range of values of F is 0 to 1.00 and of D -1.00 to $+1.00$. The higher the facility index the easier the item; the higher the discrimination index the greater the tendency for candidates doing well on that item to do well on the paper overall. The items used in these papers have a wide range of F values, each test including some relatively easy items, and, in general, positive D values.

Apart from their obvious use as revision material it is hoped that these papers will be used for classroom assessment as each topic is completed. The keys, together with facility indices, are given at the end of the book and may easily be removed if required. SI units[b] and IUPAC nomenclature[c] as recommended by the Association for Science Education are used throughout; the data needed for answering items marked with an asterisk can be found on pages xii–xiii.

---

[b] *SI Units, Signs, Symbols and Abbreviations* (The Association for Science Education)
[c] *Chemical Nomenclature, Symbols and Terminology* (The Association for Science Education)

To the student

There are 15 multiple choice test papers in this book, each consisting of 20 items, together with a final revision test of 50 items. You should be able to complete each of the papers 1–15 in a single 40-minute class period or a similar homework period and the revision test 16 in $1\frac{1}{4}$ hours. You should try to answer *all* the items in the test paper; there are five alternative answers (*responses*) for each, only one of which is correct. Four different types of item are used, directions for which can be found on page xi. Read each item carefully, then decide on the best answer and mark your choice in pencil on the answer sheet provided by your teacher (see page x). You should also have some paper for rough work available. Do not mark more than one letter for a particular item.

Some items are designed to test your ability to recall knowledge, some to test your understanding of chemical principles and some to test your ability to apply this understanding in different situations.

The data you need for answering items marked with an asterisk (*) can be found on pages xii–xiii.

Acknowledgements

The author and publisher wish to express their thanks to Dr K. F. Reid and the A-level chemistry pupils at Methodist College, Belfast, and to Mr James King and the A-level chemistry pupils at Thomas Bennett School, Crawley, for taking part in the pretesting of these multiple choice test papers, and to all those who have helped with the scrutiny of the items.

For further reading

C. V. T. Campbell and W. J. Milne, *The Principles of Objective Testing in Chemistry* (Heinemann).
E. W. Jenkins, *Objective Testing: A Guide for Science Teachers* (Centre for Studies in Science Education, Leeds University).
H. G. Macintosh and R. B. Morrison, *Objective Testing* (University of London Press).
J. C. Mathews, *Nuffield Advanced Science (Chemistry): Examinations and Assessment* (Penguin Books).

Specimen answer sheet

Note Tests 1–15 include 20 items in each test, but the final revision test includes 50 items.

Name .... Test number ....

1	~~A~~	B	C	D	E	11	A	B	C	~~D~~	E
2	A	B	~~C~~	D	E	12	A	B	C	D	~~E~~
3	A	B	~~C~~	D	E	13	A	B	C	~~D~~	E
4	A	B	C	D	~~E~~	14	A	B	~~C~~	D	E
5	A	B	C	~~D~~	E	15	A	~~B~~	C	D	E
6	A	B	~~C~~	D	E	16	~~A~~	B	C	D	E
7	A	~~B~~	C	D	E	17	A	~~B~~	C	D	E
8	~~A~~	B	C	D	E	18	A	B	~~C~~	D	E
9	A	~~B~~	C	D	E	19	A	B	C	~~D~~	E
10	A	B	~~C~~	D	E	20	A	B	C	D	~~E~~

Total score ....

Directions for answering the different types of item

1 MULTIPLE CHOICE ITEMS

Only ONE of the five responses lettered **A–E** is correct. Mark your choice on the answer sheet provided.

2 MULTIPLE COMPLETION ITEMS

ONE or MORE THAN ONE of the four responses numbered **1–4** may be correct. Consider each of the responses carefully and decide whether or not it is correct. Mark your answer sheet as follows.

A	B	C	D	E
only 1, 2 and 3 correct	only 1 and 3 correct	only 2 and 4 correct	only 4 correct	some other response, or combination of responses, correct

3 CLASSIFICATION AND SITUATION SETS

(*a*) *Classification sets*. Five different types of compound or chemical reaction or five categories of almost any kind are given for a set of items.
(*b*) *Situation sets*. A description of a particular experiment or some experimental data is followed by a set of items.

Both (*a*) and (*b*) are similar in form to multiple choice items (**1**). In each set choose the response **A, B, C, D** or **E** that correctly answers each of the items in the set. Each letter may be used ONCE, MORE THAN ONCE or NOT AT ALL.

4 ASSERTION–REASON ITEMS

A statement (*assertion*) is followed by a *reason*. Consider the assertion on its own and decide whether it is a true statement. Then consider the reason on its own and decide whether it is a true statement. If you decide that BOTH the assertion AND the reason are true, consider whether the reason is a correct explanation of the assertion. Mark your answer sheet as follows.

	Assertion	*Reason*	
A	True	True	Reason is a CORRECT EXPLANATION of the assertion
B	True	True	Reason is NOT A CORRECT EXPLANATION of the assertion
C	True	False	
D	False	True	
E	False	False	

Data

Note Items which may involve the use of some of the following data are indicated thus *.

1 APPROXIMATE RELATIVE ATOMIC MASSES

| H | 1.0 | C | 12 | N | 14 | O | 16 |
| S | 32 | Cl | 35.5 | Zn | 65 | Pb | 207 |

2 ACCURATE RELATIVE ATOMIC MASSES

H 1.008   C 12.011   N 14.007   O 15.999

3 PHYSICAL CONSTANTS

Avogadro constant	L	6.02×10^{23} mol^{-1}
Faraday constant	F	9.65×10^{4} C mol^{-1}
Gas constant	R	8.31 J K^{-1} mol^{-1}
Ice point temperature	T_{ice}	273 K
Molar volume of ideal gas at s.t.p.^{a}	$V_m^{\ominus}$	2.24×10^{-2} m^3 mol^{-1}

^{a} 273 K (0 °C) and 101 kPa (760 mmHg)

4 IONIC RADII

| Li$^+$ | Be^{2+} | B^{3+} |
| 0.060 nm | 0.031 nm | 0.020 nm |

| Na$^+$ | Mg^{2+} | Al^{3+} |
| 0.095 nm | 0.065 nm | 0.050 nm |

| K$^+$ | Ca^{2+} |
| 0.133 nm | 0.099 nm |

5 ADDITIONAL ATOMIC NUMBERS

90 thorium (Th)   91 protoactinium (Pa)   92 uranium (U)
93 neptunium (Np)   94 plutonium (Pu)

6 PERIODIC TABLE OF THE ELEMENTS WITH ATOMIC NUMBERS

1 H hydrogen																	2 He helium
3 Li lithium	4 Be beryllium											5 B boron	6 C carbon	7 N nitrogen	8 O oxygen	9 F fluorine	10 Ne neon
11 Na sodium	12 Mg magnesium											13 Al aluminium	14 Si silicon	15 P phosphorus	16 S sulphur	17 Cl chlorine	18 Ar argon
19 K potassium	20[a] Ca calcium	21 Sc scandium	22 Ti titanium	23 V vanadium	24 Cr chromium	25 Mn manganese	26 Fe iron	27 Co cobalt	28 Ni nickel	29 Cu copper	30 Zn zinc	31 Ga gallium	32 Ge germanium	33 As arsenic	34 Se selenium	35 Br bromine	36 Kr krypton
37 Rb rubidium	38 Sr strontium															53 I iodine	54 Xe xenon
55 Cs caesium	56 Ba barium																

[a] Ten d-block elements: 21 Sc scandium ...

1 Atomic structure

1. The relative atomic masses of the elements are not usually whole numbers because
   - **A** methods for determining relative atomic masses are not very accurate.
   - **B** it is often impossible to obtain a pure sample of an element.
   - **C** the mass of an electron, though relatively small, cannot be neglected.
   - **D** the number of neutrons in an atom often differs from the number of protons.
   - **E** most elements occur naturally as a mixture of isotopes.

2. The diagram below represents the mass spectrum of the element cerium.

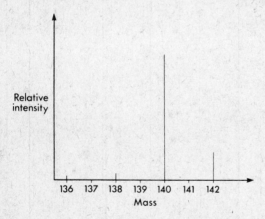

The most likely value for the relative atomic mass of cerium is
   - **A** 139.00
   - **B** 139.88
   - **C** 140.00
   - **D** 140.12
   - **E** 141.00

1 Atomic structure

3* A sample of a pure compound in a mass spectrometer is found to have a relative molecular mass of 44.053. Which one of the following molecular formulae could correctly represent this compound?

  A C_3H_8
  B CO_2
  C N_2O
  D C_2H_4O
  E C_2H_6N

4 A thin piece of metal foil is bombarded with a fast-moving stream of positively-charged particles. Which of the diagrams below most accurately represents the observed path of the particles?

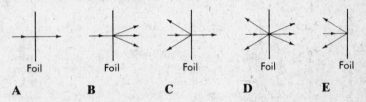

A   B   C   D   E

5 The particle X in the nuclear reaction represented by the equation

$$^{9}_{4}Be + X \rightarrow {}^{12}_{6}C + {}^{1}_{0}n$$

is

  A   an α-particle
  B   a β-particle
  C   a neon ion
  D   a deuteron (^2H)
  E   a proton

6 The product obtained by the loss of a β-particle from an atom of the isotope $^{32}_{15}P$ can be represented by

  A $^{28}_{13}Al$
  B $^{36}_{17}Cl$
  C $^{28}_{15}P$
  D $^{32}_{16}S$
  E $^{32}_{14}Si$

7 The radioactive decay of $^{238}_{92}U$ to $^{206}_{82}Pb$ involves the loss of

  A   2 α- and 8 β-particles
  B   6 α- and 8 β-particles
  C   8 α- and 2 β-particles
  D   8 α- and 6 β-particles
  E   8 α- and 10 β-particles

1 Atomic structure

8 The half-life of the isotope astatine-211 is 7.5 hours. What fraction of the original astatine will remain after 30 hours?

    A 1/32
    B 1/16
    C 1/8
    D 1/4
    E none of these

9 The decay curve of an isotope of the element lawrencium (Lr) is shown in the diagram below.

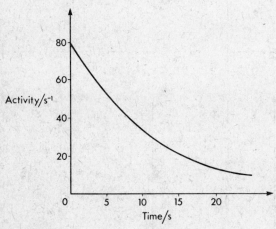

The half-life of this isotope is therefore

    A 4 s
    B 8 s
    C 16 s
    D 24 s
    E none of these

10 Which one of the following descriptions of radiocarbon dating is *not* correct?

    A ^{14}C is produced by the action of cosmic radiation on nitrogen in the atmosphere.
    B ^{14}C is oxidised to $^{14}CO_2$ which mixes with non-radioactive CO_2 in the atmosphere.
    C A living organism contains the same proportion of ^{14}C as there is in the atmosphere.
    D When a living organism dies its ^{14}C decays by the reaction $^{14}C \rightarrow {}^{14}N + \beta$.
    E The method can only be used for the determination of ages greater than the half-life of ^{14}C.

1 Atomic structure

11 Which one of the diagrams below most accurately represents the relative energies of the different orbitals in a hydrogen atom?

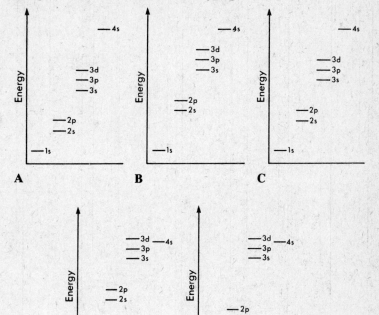

12 The diagram (below) represents the boundary surface of which one of the following types of orbital?

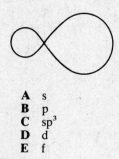

- **A** s
- **B** p
- **C** sp^3
- **D** d
- **E** f

1 Atomic structure

13 Which one of the diagrams below most accurately represents the atomic emission spectrum of hydrogen in the visible region?

```
  A           B           C           D           E
              Increasing frequency ──▶
```

MULTIPLE COMPLETION ITEMS

ONE or MORE THAN ONE of the four responses numbered 1–4 may be correct. Consider each of the responses carefully and decide whether or not it is correct. Mark your answer sheet as follows.

A	B	C	D	E
only 1, 2 and 3 correct	only 1 and 3 correct	only 2 and 4 correct	only 4 correct	some other response, or combination of responses, correct

14 In the X-ray spectra of cobalt and nickel the frequencies of corresponding lines for nickel are higher than those for cobalt. It can be deduced *from this observation alone* that

1. the atomic number of nickel is higher than that of cobalt.
2. a nickel atom has more extranuclear electrons than a cobalt atom.
3. nickel must follow cobalt in the periodic table of the elements.
4. the relative atomic mass of nickel is higher than that of cobalt.

15 Natural gallium (atomic number 31, relative atomic mass 69.7) consists of a mixture of two isotopes of mass number 69 and 71. Which of the following conclusions can be drawn from these data?

1. Gallium-69 is more abundant than gallium-71.
2. There are 38 and 40 neutrons respectively in an atom of the two isotopes.
3. An atom of either isotope has the electron configuration 2.8.18.3.
4. Gallium-71 is radioactive.

16 Which of the following statements relating to the nuclear fission of uranium-238 is/are correct?

1. The nucleus is bombarded with high-speed protons.
2. Several neutrons are released as a result of the fission process.
3. The fission product is an isotope of an element with an atomic number higher than that of uranium.
4. A chain reaction may occur.

1 Atomic structure

17 The mass spectrum of bromine vapour consists of only three peaks at mass 158, 160 and 162. Which of the following conclusions can be drawn from this observation?
    1 Natural bromine is a mixture of three isotopes.
    2 The peak at mass 158 is due to $^{79}Br_2^+$.
    3 The peak at mass 160 is due to $^{80}Br_2^+$.
    4 The peak at mass 162 is due to $^{81}Br_2^+$.

18* The mass spectra of two alkanes **X** and **Y** are shown in the diagrams below.

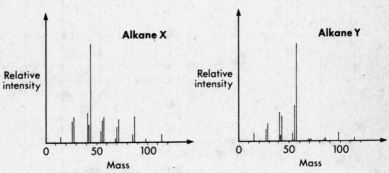

It can be deduced from these mass spectra that
    1 the two alkanes are isomeric.
    2 the relative molecular mass of alkane **X** is 43.
    3 the line at mass 15 in both spectra is due to the ion CH_3^+.
    4 the molecular formula of alkane **Y** is C_4H_{10}.

19 Which of the following statements about the atomic emission spectrum of hydrogen is/are correct?
    1 The frequency v of each line depends on the difference in energy between the higher and lower levels (ΔE) according to the equation $\Delta E = hv$.
    2 The spectrum consists of several series of lines corresponding to electronic transitions to the levels $n = 1, 2, 3, \ldots$
    3 Electronic transitions to the level $n = 2$ give rise to lines in the visible region.
    4 The energy corresponding to the convergence limit of the Balmer series of lines (for which $n = 2$) is the ionisation energy of hydrogen.

1 Atomic structure

20 The diagram (right) could represent the boundary surface of
  1. a p_x orbital
  2. a p_y orbital
  3. a p_z orbital
  4. an s orbital

2 Structure and bonding

1. Which one of the following properties of a solid provides the most conclusive evidence for the presence of ionic (electrovalent) bonds?
   - **A** crystalline
   - **B** soluble in water
   - **C** high melting point
   - **D** conducts electricity when molten
   - **E** conducts electricity when dissolved in water

2. The melting point, boiling point and molar conductivity Λ (in the molten state) of the chlorides of three elements E_1, E_2 and E_3 are given in the table below.

Chloride	m.p./K	b.p./K	$\Lambda/\text{S cm}^2 \text{ (mol of Cl)}^{-1}$
of E_1	883	1 650	175
of E_2	1 148	2 750	62
of E_3	548	1 005	0.02

   Which one of the following conclusions can be drawn from these data?
   - **A** All three chlorides are ionic (electrovalent).
   - **B** Only the chlorides of E_1 and E_2 are ionic.
   - **C** Only the chloride of E_1 is ionic.
   - **D** Only the chloride of E_2 is ionic.
   - **E** None of the three chlorides is ionic.

3. Which one of the following anhydrous chlorides is *not* hydrolysed in water to give hydrogen chloride?
   - **A** $AlCl_3$
   - **B** $MgCl_2$
   - **C** PCl_3
   - **D** SCl_2
   - **E** $SiCl_4$

2 Structure and bonding

4 Which one of the following substances in the solid state consists of atoms or molecules held together mainly by van der Waals forces?
- **A** aluminium
- **B** phenol
- **C** silicon(IV) oxide
- **D** tetrachloromethane
- **E** water

5 With how many other carbon atoms is each carbon atom in graphite covalently bonded?
- **A** 1
- **B** 2
- **C** 3
- **D** 4
- **E** 8

6 Which one of the diagrams below represents the most likely shape of the phosphine molecule ($PH_3(g)$)?

- **A** H—P—H, | H (T-shaped)
- **B** P H / \ / H H (non-linear)
- **C** P—H | | H—H (square planar)
- **D** P / | \ H H H (pyramidal)
- **E** H \ P—H / H (trigonal planar)

7 Which one of the following types of bonding best describes the hydration of ions in aqueous solution?
- **A** coordinate (dative covalent)
- **B** covalent
- **C** dipole–dipole
- **D** hydrogen
- **E** ion–dipole

8* For which one of the following ions is the hydration energy greatest?
- **A** Al^{3+}
- **B** Ca^{2+}
- **C** Li^+
- **D** Mg^{2+}
- **E** Na^+

2 Structure and bonding

9 The S—O bond length in the sulphate(VI) ion is 0.151 nm. Which one of the following is the best representation of the structure of this ion? (Calculated bond lengths: sulphur–oxygen single bond 0.178 nm, sulphur–oxygen double bond 0.149 nm.)

A $\left[\begin{array}{c} O \\ \uparrow \\ O \leftarrow S \rightarrow O \\ \downarrow \\ O \end{array} \right]^{2-}$
B $\left[\begin{array}{c} O \\ | \\ O=S-O \\ || \\ O \end{array} \right]^{2-}$
C $\left[\begin{array}{c} O \\ | \\ O-S-O \\ | \\ O \end{array} \right]^{2-}$

D $\left[\begin{array}{c} O \\ || \\ O=S=O \\ || \\ O \end{array} \right]^{2-}$
E $\left[\begin{array}{c} O \\ \vdots \\ O=S=O \\ \vdots \\ O \end{array} \right]^{2-}$

10 When a halogenoalkane RX reacts with magnesium in a solvent such as ethoxyethane ('ether') a so-called Grignard reagent is obtained which is often represented as RMgX. The solution conducts electricity and magnesium moves to both the anode and the cathode. On the basis of this evidence alone which one of the following is the best representation of the situation in the Grignard solution?

   A RMgX
   B $RMg^+ X^-$
   C $R^+ MgX^-$
   D $R_2Mg + MgX_2$
   E $RMg^+ + RMgX_2^-$

CLASSIFICATION SET

Items 11 to 13

The melting point, boiling point and electrical conductivity of five different substances are given in the table below.

Substance	m.p./K	b.p./K	Electrical conductivity	
			Solid	Liquid
A	234	630	Good	Good
B	720	1 450	Poor	Good
C	279	353	Poor	Poor
D	2 000	2 500	Poor	Poor
E	453	1 600	Good	Good

Select the lettered substance **A–E** that fits the description given in each of the following items. Each letter may be used ONCE, MORE THAN ONCE or NOT AT ALL.

2 Structure and bonding

11 An alkali metal.

12 A covalent compound with discrete molecules.

13 A covalent compound with a macromolecular structure.

MULTIPLE COMPLETION ITEMS

ONE or MORE THAN ONE of the four responses numbered 1–4 may be correct. Consider each of the responses carefully and decide whether or not it is correct. Mark your answer sheet as follows.

A	B	C	D	E
only 1, 2 and 3 correct	only 1 and 3 correct	only 2 and 4 correct	only 4 correct	some other response, or combination of responses, correct

14 Which of the following types of bond is/are found in solid hydrated copper(II) sulphate(VI), $CuSO_4.5H_2O$?
  1. covalent
  2. hydrogen
  3. ionic (electrovalent)
  4. metallic

15 The structure of the hydrogenfluoride ion in the salt potassium hydrogenfluoride may be correctly represented as ($\cdots$ denotes a hydrogen bond)
  1. $[F \rightarrow H - F]^-$
  2. $[F \cdots H - F]^-$
  3. $[F - H \leftarrow F]^-$
  4. $[F - H \cdots F]^-$

16 Which of the species represented by the following formulae is/are tetrahedral in shape?
  1. BF_4^-
  2. CF_4
  3. PF_4^+
  4. SF_4

17 The diagram (right) could represent a molecule of
  1. CO_2
  2. CS_2
  3. N_2O
  4. SO_2

2 Structure and bonding

18 The diagrams below show the electron configuration of a carbon atom in the ground state (I), in the excited state (II) and in the sp^3 hybridised state (III).

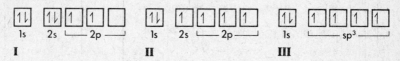

Which of the following statements is/are correct?
1. In the ground state a carbon atom can form only two covalent bonds.
2. In the excited state a carbon atom can form four covalent bonds.
3. In the excited state a carbon atom forms identical covalent bonds.
4. In the sp^3 hybridised state a carbon atom can form four identical covalent bonds.

19 In which of the following elements can the bonding electrons be considered to be delocalised?
1. copper
2. diamond
3. graphite
4. sulphur

20* In which of the following experimental observations is hydrogen bonding likely to be a relevant factor?
1. The deflection of a jet of ethanol in an electrostatic field.
2. The relative molecular mass of ethanoic acid in benzene solution is approximately 120.
3. The boiling point of 1-chlorobutane ($CH_3CH_2CH_2CH_2Cl$) is higher than that of 2-chloro-2-methylpropane (($CH_3)_3CCl$).

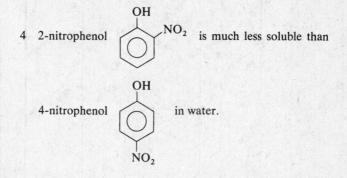

3 The periodic table

1

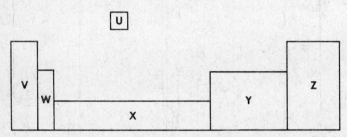

In the outline extended form of the periodic table shown above, in which one of the following areas are the p-block elements to be found?

A   U
B   U+V
C   W+Y
D   X
E   Z

2 Which one of the following elements is represented in the incomplete oxidation number diagram shown right? (X is the symbol for the element.)

A   carbon
B   chlorine
C   nitrogen
D   sulphur
E   xenon

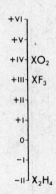

3 The periodic table

3 A solid element (symbol Y) conducts electricity and forms two chlorides YCl_n (a colourless, volatile liquid) and YCl_{n-2} (a colourless solid). To which one of the following groups in the periodic table does Y most probably belong?

- **A** III
- **B** IV
- **C** V
- **D** VI
- **E** d-block (transition) element

4 In which group of the periodic table would you expect to find the element **Z** on the graph of first ionisation energy and atomic number below?

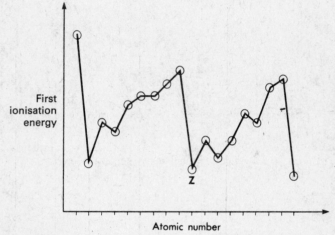

- **A** I
- **B** II
- **C** IV
- **D** VII
- **E** VIII (noble gas)

5 In which group of the periodic table would you expect to find an element which has successive ionisation energies of 940 (1st), 2 080, 3 090, 4 140, 7 030, 7 870, 16 000 and 19 500 kJ mol^{-1}?

- **A** I
- **B** IV
- **C** V
- **D** VI
- **E** VII

3 The periodic table

6 In which group of the periodic table would you expect to find the element **X** on the graph of boiling point and atomic number below? (The eight consecutive elements shown all have atomic number ≤ 20.)

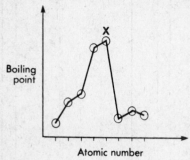

- **A** I
- **B** III
- **C** IV
- **D** V
- **E** VI

CLASSIFICATION SETS

Items 7 to 10

The electron configurations of five different elements are given in the table below.

Element	Electron configuration
A	2.8.3
B	2.8.5
C	2.8.13.2
D	2.8.18.8.2
E	2.8.18.18.8

Select the lettered electron configuration from **A–E** that correctly represents each of the following elements. Each letter may be used ONCE, MORE THAN ONCE or NOT AT ALL.

7 A noble gas.

8 An alkaline earth metal.

9 A d-block (transition) element.

10 Forms an ion with a −3 charge.

3 The periodic table

Items 11 to 13

The graph below shows the variation in the molar enthalpy change of vaporisation ΔH_v with atomic number Z for eight consecutive elements in the periodic table all with atomic number ≤ 20.

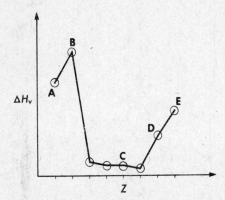

Select the lettered element **A–E** that fits the description given in each of the following items. Each letter may be used ONCE, MORE THAN ONCE or NOT AT ALL.

11 The element that is most likely to be in Group I of the periodic table.

12 The element that is most likely to consist of diatomic molecules.

13 The element that is likely to have the highest boiling point.

Items 14 to 17

The graphs opposite show the variation in five different physical properties of various elements and their compounds.

Select the lettered graph **A–E** that correctly represents the physical properties in each of the following items. Each letter may be used ONCE, MORE THAN ONCE or NOT AT ALL.

14 Ionic radii of F^-, Cl^-, Br^- and I^-.

15 First ionisation energies of Li, Na, K and Rb.

3 The periodic table

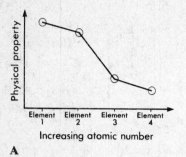

A

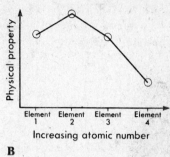

B

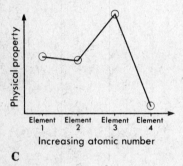

C

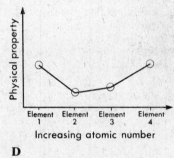

D

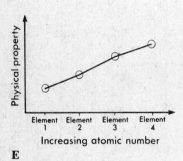

E

16 Boiling points of the hydrides of nitrogen (NH_3), phosphorus (PH_3), arsenic (AsH_3) and antimony (SbH_3).

17 Melting points of Mg, Al, Si and P.

3 The periodic table

MULTIPLE COMPLETION ITEMS

ONE or MORE THAN ONE of the four responses numbered 1–4 may be correct. Consider each of the responses carefully and decide whether or not it is correct. Mark your answer sheet as follows.

A	B	C	D	E
only 1, 2 and 3 correct	only 1 and 3 correct	only 2 and 4 correct	only 4 correct	some other response, or combination of responses, correct

18 Which of the following properties of the elements vary in a periodic way with atomic number?

  1 electrical conductivity
  2 electronegativity
  3 enthalpy change of vaporisation
  4 first ionisation energy

19 Which of the following properties is/are characteristic of the elements carbon, silicon, tin and lead in Group IV of the periodic table? (X denotes any of these Group IV elements.)

  1 Giant lattice structure.
  2 Form an oxide XO_2.
  3 Form a covalent chloride XCl_4.
  4 Form a complex ion $[XHal_6]^{2-}$, where Hal denotes a halogen.

20 An element has the following properties:

  electron configuration 2.8.18.18.7
  melting point 114 °C
  boiling point 184 °C

  Which of the following conclusions about the element can be drawn from these data?

  1 It is in Group VII of the periodic table.
  2 It has a giant covalent lattice.
  3 It has an oxidation number of $-I$.
  4 It is bromine.

4 States of matter

1* Plugs of cotton wool soaked in concentrated hydrochloric acid and concentrated ammonia solution are simultaneously placed in the ends of a long, horizontal glass tube and rubber bungs are then inserted in the ends. The apparatus is left for some time when it is found that a white solid forms at **X** in the tube. Which one of the diagrams below most accurately represents the situation?

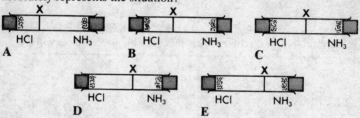

2* A plastic bottle is weighed full of oxygen and then full of a gas **X**. The mass of oxygen in the bottle is 0.64 g and the mass of gas **X** is 0.52 g. What is the relative molecular mass of **X**?

    A   13
    B   26
    C   52
    D   104
    E   208

3 50 cm^3 of the gaseous hydride of an element decomposes on heating to give the element and 100 cm^3 of hydrogen gas, measured at the same temperature and pressure. How many hydrogen atoms does one molecule of the hydride contain?

    A   1
    B   2
    C   3
    D   4
    E   8

4 States of matter

4* Which one of the following values correctly represents the volume of 1 mol of an ideal gas at room temperature (15 °C) and a pressure of 100 kPa (750 mmHg)?

- **A** 21.0 dm^3
- **B** 21.5 dm^3
- **C** 22.4 dm^3
- **D** 24.0 dm^3
- **E** 44.8 dm^3

5* A certain mass of a gas of relative molecular mass 30 is found to occupy a volume of 50.0 cm^3 at 27 °C and 100 kPa (750 mmHg) pressure. What is the mass of the gas?

- **A** 0.000667 g
- **B** 0.00600 g
- **C** 0.0600 g
- **D** 0.667 g
- **E** 6.00 g

6 Which one of the diagrams below most correctly represents the distribution of molecular speeds in a gas at a given temperature?

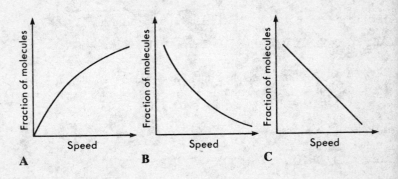
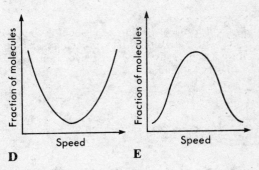

4 States of matter

7 The diagram below represents the unit cell of the crystal structure of the compound formed by a metal (symbol M) with a non-metal (symbol X). (• represents a metal ion and o a non-metal ion.)

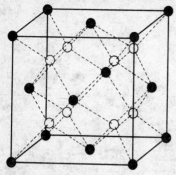

The formula of the compound is

- A M_2X
- B MX
- C MX_2
- D MX_3
- E MX_4

CLASSIFICATION SET

Items 8 to 11

Five physical methods used for the determination of atomic and molecular structure are given below.

- A electron diffraction
- B infrared absorption spectroscopy
- C mass spectrometry
- D X-ray diffraction
- E X-ray spectroscopy

Select the lettered method **A–E** that you would use for the determination of the physical quantities in each of the following items. Each letter may be used ONCE, MORE THAN ONCE or NOT AT ALL.

8 The relative molecular mass of an organic compound.

9 The O—H bond length in water.

10 The Cl—Cl bond length in chlorine gas.

11 The C—C bond length in diamond.

4 States of matter

MULTIPLE COMPLETION ITEMS

ONE or MORE THAN ONE of the four responses numbered 1–4 may be correct. Consider each of the responses carefully and decide whether or not it is correct. Mark your answer sheet as follows.

A	B	C	D	E
only 1, 2 and 3 correct	only 1 and 3 correct	only 2 and 4 correct	only 4 correct	some other response, or combination of responses, correct

12 Which of the following equations is/are correct statements of the ideal gas equation? (p denotes pressure, V volume, ρ density, T temperature, m mass of gas, M_r relative molecular mass of gas, n amount of gas and R the gas constant.)
 1 $pM_r = \rho RT$
 2 $pm = \rho RT$
 3 $pV = nRT$
 4 $pVm = M_r RT$

13 The variation of volume V with pressure p for a gas **X** at constant temperature is shown in the diagram below.

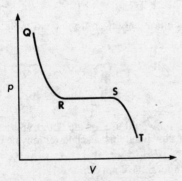

Which of the following statements is/are correct?
 1 **X** is an ideal gas.
 2 **X** is at a temperature below its critical temperature.
 3 **QR** represents the effect of pressure on the volume of gaseous **X**.
 4 **X** is undergoing liquefaction along **RS**.

4 States of matter

14 Which of the following properties would you expect a colloidal solution to possess?

1. The colloidal particles all move in the same direction in an electric field.
2. Coagulation occurs if an electrolyte is added.
3. Low osmotic pressure.
4. The colloidal particles diffuse more rapidly than the particles in a true solution.

15 Which of the following statements about the arrangement of identical particles in the structure shown right is/are correct?

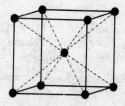

1. It is a body-centred cubic structure.
2. It is a close-packed structure.
3. The coordination number for each particle is 8.
4. It could represent caesium chloride.

16 Which of the following substances have infrared vibrational spectra?

1. HCl
2. CH_3Cl
3. ICl
4. I_2

Test Paper 4 continued overleaf

4 States of matter

ASSERTION–REASON ITEMS

A statement (*assertion*) is followed by a *reason*. Consider the assertion on its own and decide whether it is a true statement. Then consider the reason on its own and decide whether it is a true statement. If you decide that BOTH the assertion AND the reason are true, consider whether the reason is a correct explanation of the assertion. Mark your answer sheet as follows.

	Assertion	Reason	
A	True	True	Reason is a CORRECT EXPLANATION of the assertion
B	True	True	Reason is NOT A CORRECT EXPLANATION of the assertion
C	True	False	
D	False	True	
E	False	False	

	Assertion		Reason
17	Real gases deviate from ideal behaviour	BECAUSE	the molecules of a gas occupy a definite volume and exert attractive forces on each other.
18	The relative molecular mass of dinitrogen tetroxide can be determined by Victor Meyer's method	BECAUSE	dinitrogen tetroxide is a highly volatile liquid.
19	$^{238}UF_6$ vapour diffuses more rapidly than $^{235}UF_6$ vapour	BECAUSE	the relative molecular mass of $^{238}UF_6$ is higher than that of $^{235}UF_6$.
20	A diffraction pattern is produced when a beam of X-rays is allowed to fall on a finely powdered crystalline substance	BECAUSE	the crystals in a fine powder are randomly orientated so that there are always some with their lattice planes in the correct positions for maximum X-ray reflection to occur.

5 Energetics

1. Which one of the following equations correctly represents the reaction to which the enthalpy change of combustion of sulphur refers?
   - **A** $S(s) + O_2(g) \rightarrow SO_2(g)$
   - **B** $S(g) + O_2(g) \rightarrow SO_2(g)$
   - **C** $S(s) + \frac{3}{2}O_2(g) \rightarrow SO_3(g)$
   - **D** $S(g) + \frac{3}{2}O_2(g) \rightarrow SO_3(g)$
   - **E** $S(s) + \frac{3}{2}O_2(g) \rightarrow SO_3(s)$

2. Which one of the following equations correctly represents the reaction to which the enthalpy change of formation of methane refers?
   - **A** $C(diamond) + 2H_2(g) \rightarrow CH_4(g)$
   - **B** $C(diamond) + 4H(g) \rightarrow CH_4(g)$
   - **C** $C(graphite) + 2H_2(g) \rightarrow CH_4(g)$
   - **D** $C(graphite) + 4H(g) \rightarrow CH_4(g)$
   - **E** $C(g) + 4H(g) \rightarrow CH_4(g)$

3. The enthalpy change of combustion (ΔH_c) for three different alcohols is given in the table below.

Alcohol	ΔH_c/kJ mol^{-1}
Methanol	−715
Ethanol	−1 370
Propan-1-ol	−2 010

   ΔH_c for pentan-1-ol is most likely to be
   - **A** $-2\,660$ kJ mol^{-1}
   - **B** $-3\,310$ kJ mol^{-1}
   - **C** $-3\,575$ kJ mol^{-1}
   - **D** $-3\,960$ kJ mol^{-1}
   - **E** impossible to predict from the available data

5 Energetics

4 The enthalpy change of combustion of graphite is -393.5 kJ mol^{-1} and that of diamond is -395.4 kJ mol^{-1}. The enthalpy change for the reaction

$$C(graphite) \rightarrow C(diamond)$$

is therefore

- **A** -788.9 kJ mol^{-1}
- **B** -1.9 kJ mol^{-1}
- **C** 0
- **D** $+1.9$ kJ mol^{-1}
- **E** $+788.9$ kJ mol^{-1}

5 The enthalpy change of atomisation of sulphur is $+224$ kJ mol^{-1}. For which one of the reactions represented by the following equations is the enthalpy change equal to $+224$ kJ?

- **A** $S(rhombic) \rightarrow S(g)$
- **B** $S_8(rhombic) \rightarrow 8S(g)$
- **C** $\frac{1}{8}S_8(rhombic) \rightarrow S(g)$
- **D** $S_8(rhombic) \rightarrow S_8(monoclinic)$
- **E** $\frac{1}{8}S_8(rhombic) \rightarrow \frac{1}{8}S_8(monoclinic)$

6 The enthalpy change for the process

$$\tfrac{1}{2}Cl_2(g) \rightarrow Cl^+(g) + e^-$$

is equal to the

- **A** electron affinity of chlorine.
- **B** first ionisation energy of chlorine.
- **C** sum of the enthalpy change of atomisation and the electron affinity of chlorine.
- **D** sum of the enthalpy change of atomisation and the first ionisation energy of chlorine.
- **E** sum of the bond dissociation energy and the electron affinity of chlorine.

7 Which one of the following equations correctly represents the reaction to which the lattice energy of calcium oxide refers?

- **A** $Ca(s) + \tfrac{1}{2}O_2(g) \rightarrow CaO(s)$
- **B** $Ca(g) + \tfrac{1}{2}O_2(g) \rightarrow CaO(s)$
- **C** $Ca^{2+}(g) + O^{2-}(g) \rightarrow CaO(s)$
- **D** $Ca^{2+}(g) + O^{2-}(g) \rightarrow CaO(g)$
- **E** $Ca^{2+}(aq) + O^{2-}(aq) \rightarrow CaO(s)$

5 Energetics

8 The enthalpy change of hydrogenation of but-1-ene is $-127\,\text{kJ mol}^{-1}$. The most likely value for the enthalpy change of hydrogenation of buta-1,3-diene ($H_2C=CH-CH=CH_2$) is

- **A** $-120\,\text{kJ mol}^{-1}$
- **B** $-127\,\text{kJ mol}^{-1}$
- **C** $-239\,\text{kJ mol}^{-1}$
- **D** $-254\,\text{kJ mol}^{-1}$
- **E** $-269\,\text{kJ mol}^{-1}$

9 The Born–Haber cycle for the dissolving of sodium fluoride in water is shown in the diagram below.

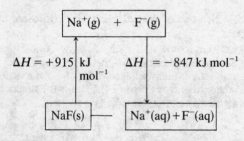

The enthalpy change of solution of sodium fluoride is

- **A** $-847\,\text{kJ mol}^{-1}$
- **B** $-68\,\text{kJ mol}^{-1}$
- **C** $+68\,\text{kJ mol}^{-1}$
- **D** $+847\,\text{kJ mol}^{-1}$
- **E** impossible to calculate from the available data

10 The enthalpy changes of atomisation of methane and ethane are 1 648 and 2 820 kJ mol^{-1} respectively. The C—C mean bond dissociation energy is therefore

- **A** $-1\,172\,\text{kJ mol}^{-1}$
- **B** $-348\,\text{kJ mol}^{-1}$
- **C** $+348\,\text{kJ mol}^{-1}$
- **D** $+1\,172\,\text{kJ mol}^{-1}$
- **E** impossible to calculate from the available data

Test paper 5 continued overleaf

5 Energetics

11 In which one of the reactions represented by the following equations is the entropy change ($\Delta S^\ominus$) likely to be positive?

- A $C_2H_5OH(g) \rightarrow C_2H_5OH(l)$
- B $N_2(g) + 3H_2(g) \rightarrow 2NH_3(g)$
- C $SO_3(g) + H_2O(l) \rightarrow H_2SO_4(l)$
- D $CuSO_4(s) + 5H_2O(l) \rightarrow CuSO_4.5H_2O(s)$
- E $CH_3CH_2CH_2CH_3(g) \rightarrow CH_3CH_3(g) + H_2C{=}CH_2(g)$

12 The standard Gibbs free energy change ($\Delta G^\ominus$) for the cell reaction

$$Zn(s) + Cu^{2+}(aq) \rightarrow Zn^{2+}(aq) + Cu(s)$$

is negative. Which one of the following conclusions can be drawn from this information?

- A The reaction is slow under standard conditions.
- B The equilibrium constant for the reaction is small.
- C The standard enthalpy change ($\Delta H^\ominus$) for the reaction is negative.
- D The standard entropy change ($\Delta S^\ominus$) for the reaction is positive.
- E The standard cell potential ($E^\ominus$) is positive.

13* The stability constant K_{stab} for the formation of the diamminecopper(I) ion

$$Cu^+ + 2NH_3 \rightleftharpoons [Cu(NH_3)_2]^+$$

is equal to 1.0×10^{11} dm^6 mol^{-2} at 25 °C. Which one of the following approximate values is correct for the standard Gibbs free energy change $\Delta G^\ominus$/kJ mol^{-1} for the reaction?

- A -63
- B -27
- C $+27$
- D $+63$
- E none of these

MULTIPLE COMPLETION ITEMS

ONE or MORE THAN ONE of the four responses numbered 1–4 may be correct. Consider each of the responses carefully and decide whether or not it is correct. Mark your answer sheet as follows.

A	B	C	D	E
only 1, 2 and 3 correct	only 1 and 3 correct	only 2 and 4 correct	only 4 correct	some other response, or combination of responses, correct

5 Energetics

14 For which of the changes represented by the following equations is the enthalpy change likely to be negative?

1 $Cl(g) + e^- \rightarrow Cl^-(g)$
2 $Cl^-(g) + e^- \rightarrow Cl^{2-}(g)$
3 $Cl^-(g) + aq \rightarrow Cl^-(aq)$
4 $Cl(g) \rightarrow Cl^+(g) + e^-$

15 Values for the enthalpy change of neutralisation (ΔH_n) for various acid–base pairs are given in the table below.

Acid	Base	ΔH_n/kJ mol^{-1}
HCl	NaOH	−57.1
HNO$_3$	NH$_3$	−52.2
HA	KOH	−55.2

The acid HA could be

1 H_2SO_4
2 CH_3COOH
3 CCl_3COOH
4 C_6H_5COOH

16 The Born–Haber cycle for the formation of potassium bromide is shown in the diagram below.

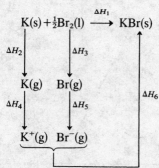

Which of the following statements is/are correct?

1 ΔH_2 represents the enthalpy change of atomisation of K.
2 ΔH_3 represents half the bond dissociation energy of Br–Br.
3 ΔH_4 represents the first ionisation energy of K.
4 ΔH_6 represents the lattice energy of KBr.

5 Energetics

17 The energy profile diagram for a reversible reaction is shown below.

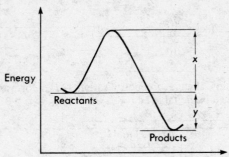

Which of the following conclusions can be drawn from this diagram?
1 The activation energy of the forward reaction is x.
2 The activation energy of the reverse reaction is y.
3 The forward reaction is exothermic.
4 The enthalpy change (ΔH) for the reverse reaction is $y - x$.

18 The energy profile diagram for the hydrolysis of a halogenoalkane is shown below.

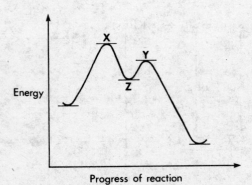

Which of the following conclusions can be drawn from this diagram?
1 The reaction is exothermic.
2 **X** and **Y** represent transition states.
3 An intermediate is formed at **Z**.
4 The first step is faster than the second.

5 Energetics

19 The variation of the standard Gibbs free energy change ($\Delta G^{\ominus}$) with temperature (T) for five different oxidation reactions is shown in the diagram below.

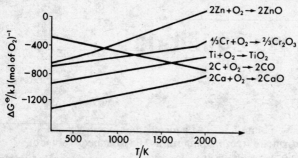

Which of the following metals could be obtained from their oxides by reduction with carbon at 1 500 K?

1 calcium
2 chromium
3 titanium
4 zinc

20 The Ellingham diagrams for four oxides of nitrogen are shown below. ($\Delta G^{\ominus}$ denotes the standard Gibbs free energy change and T the temperature.)

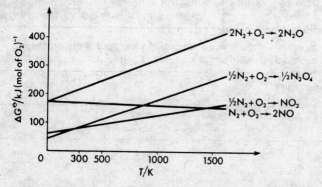

Which of the following conclusions can be drawn from these diagrams?

1 The formation of NO from its elements is energetically favoured by higher temperatures.
2 At 500 K, NO_2 is energetically more stable than its dimer N_2O_4.
3 At high temperatures N_2O tends to disproportionate to N_2 and NO.
4 At approximately 300 K the equilibrium constant for the reaction $\frac{1}{2}N_2O_4 \rightleftharpoons NO_2$ is equal to zero.

6 Phase equilibria

1. The freezing point of a 0.1 M aqueous solution of glucose is $-0.18\,°C$. The approximate freezing point of a 0.05 M aqueous solution of magnesium chloride is therefore
   A $-0.09\,°C$
   B $-0.18\,°C$
   C $-0.27\,°C$
   D $-0.36\,°C$
   E $-0.54\,°C$

2. Two aqueous solutions S_1 and S_2 of different solutes have the same vapour pressure at a given temperature. Which one of the following conclusions *cannot* be drawn from this information?
   A S_1 and S_2 contain equal masses of solute per unit volume.
   B S_1 and S_2 contain equal amounts of solute per unit volume.
   C S_1 and S_2 have the same osmotic pressure at the given temperature.
   D S_1 and S_2 have the same freezing point.
   E S_1 and S_2 have the same boiling point.

3. Which one of the following aqueous solutions has the lowest freezing point?
   A   0.1 M ethanoic acid
   B   0.1 M glucose
   C   0.1 M iron(II) chloride
   D   0.1 M sodium chloride
   E   pure water

6 Phase equilibria

4 Which one of the following methods is the most suitable for the determination of the relative molecular mass of a polymer?

- **A** depression of the freezing point of a suitable solvent
- **B** elevation of the boiling point of a suitable solvent
- **C** osmotic pressure in a suitable solvent
- **D** vapour pressure lowering of a suitable solvent
- **E** Victor Meyer (volume of air displaced by a known mass of polymer vapour)

5 The transition temperature for rhombic sulphur to monoclinic sulphur can be determined by measuring the volume of a given mass of sulphur (V_S) at various temperatures (T). Which one of the graphs below most accurately represents the change in volume with temperature? (The density of rhombic sulphur is higher than that of monoclinic sulphur.)

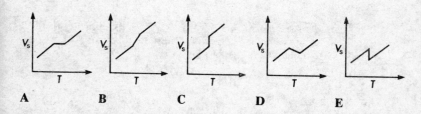

A   B   C   D   E

6 Which one of the following is *not* a correct statement about white phosphorus?

- **A** Its molecular formula is P_4.
- **B** The molecule is tetrahedral.
- **C** It is the stable form at room temperature.
- **D** Its conversion to red phosphorus is an exothermic process.
- **E** It is soluble in carbon disulphide.

7 Which one of the following equations correctly represents Raoult's law? (p and p_0 denote pressure, V volume, T temperature, m mass, n and n_0 amount of substance, R the gas constant, E activation energy, $\bar{c}$ mean speed, K_p equilibrium constant in terms of partial pressures, ΔH enthalpy change and k rate constant; A is a constant.)

- **A** $k = A\mathrm{e}^{-E/RT}$
- **B** $pV = \frac{1}{3}nm\bar{c}^2$
- **C** $pV = nRT$
- **D** $\dfrac{\mathrm{d}(\lg K_p)}{\mathrm{d}T} = \dfrac{-\Delta H}{RT^2}$
- **E** $\dfrac{p_0 - p}{p_0} = \dfrac{n}{n + n_0}$

6 Phase equilibria

8 The temperature–composition diagram for a mixture of two liquids L_1 and L_2 is shown below.

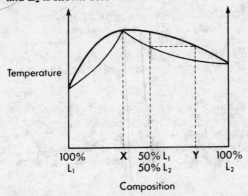

If a mixture of composition 50% L_1/50% L_2 is fractionally distilled, which one of the following correctly represents the composition of the liquid finally remaining in the flask?

- **A** 100% L_1
- **B** 100% L_2
- **C** 50% L_1/50% L_2
- **D** that given by **X** on the diagram
- **E** that given by **Y** on the diagram

SITUATION SET

Select the response from **A–E** that correctly answers each of the items in the set. Each letter may be used ONCE, MORE THAN ONCE or NOT AT ALL.

Items 9 to 11

The phase diagram for an unknown substance is shown below.

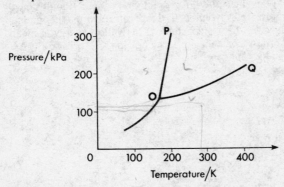

6 Phase equilibria

9 Which one of the following is represented by the point **O** on the diagram?

   A the boiling point of the liquid
   B the melting point of the solid
   C the triple point
   D the transition point from one polymorph to another
   E the critical point

10* Which one of the following correctly describes the situation for the substance at standard temperature and pressure?

   A gas phase only present
   B liquid phase only present
   C solid phase only present
   D gas and liquid in equilibrium
   E liquid and solid in equilibrium

11 Which one of the following conclusions can be drawn from the phase diagram?

   A The melting point of the solid decreases as the pressure rises.
   B The solid sublimes at standard pressure.
   C Curve **OP** represents the variation of the transition temperature with pressure.
   D Curve **OQ** represents the variation of the vapour pressure of the solid with temperature.
   E At the end of the curve **OQ** the solid and gas phases become indistinguishable.

MULTIPLE COMPLETION ITEMS

ONE or MORE THAN ONE of the four responses numbered 1–4 may be correct. Consider each of the responses carefully and decide whether or not it is correct. Mark your answer sheet as follows.

A	B	C	D	E
only 1, 2 and 3 correct	only 1 and 3 correct	only 2 and 4 correct	only 4 correct	some other response, or combination of responses, correct

12 Which of the following factors increase the proportion of liquid L_1 in the distillate when a mixture of two liquids L_1 and L_2 is steam distilled?

   1 high relative molecular mass of L_1
   2 high relative molecular mass of L_2
   3 high vapour pressure of L_1
   4 high vapour pressure of L_2

6 Phase equilibria

13 Which of the following properties make ethoxyethane ('ether') a suitable solvent for the extraction of organic substances from aqueous solution?
1. low boiling point
2. immiscibility with water
3. chemical unreactivity
4. low relative molecular mass

14 The ratio

concentration of **S** in solvent 1/concentration of **S** in solvent 2

for the distribution of a solid **S** between two immiscible solvents 1 and 2 is constant only if
1. the temperature is constant.
2. the volumes of the two solvents 1 and 2 are equal.
3. the solid **S** does not associate or dissociate in either of the two solvents 1 and 2.
4. the total mass of solid **S** present is constant.

15 The vapour pressure–temperature diagram for a pure solvent and a solution of a non-volatile solute in the solvent is shown below.

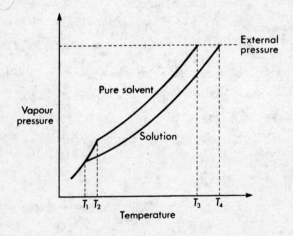

Which of the following conclusions can be drawn from this diagram?
1. The vapour pressure of a pure solvent increases as the temperature rises.
2. The pure solvent freezes at temperature T_2.
3. The solution freezes at temperature T_1.
4. The solute produces an elevation in the boiling point of the solvent of $T_4 - T_2$.

6 Phase equilibria

16 The vapour pressure–composition diagrams for four different pairs of liquids are shown below.

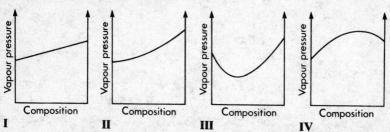

In which of these diagrams do the liquids concerned show a positive deviation from Raoult's law?
1 I
2 II
3 III
4 IV

17 The vapour pressure–temperature curves for white, red and liquid phosphorus are shown in the diagram below.

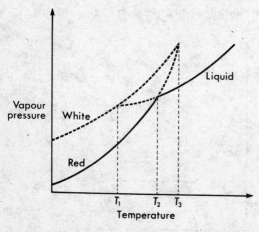

Which of the following conclusions can be drawn from this diagram?
1 White phosphorus tends to change to red phosphorus at all temperatures.
2 If liquid phosphorus is cooled rapidly, white phosphorus is formed at temperature T_1.
3 If red phosphorus is heated it melts at temperature T_2.
4 If white phosphorus is heated rapidly it melts at temperature T_3.

37

6 Phase equilibria

18 The phase diagram for the tin–lead system is shown below.

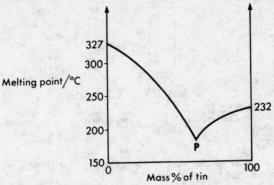

Which of the following statements is/are correct?
1. The melting point of pure tin is 232 °C.
2. The addition of tin to lead does not affect the melting point.
3. **P** is the eutectic point.
4. At **P** two phases are in equilibrium.

19 The cooling curve for a 1:1 mixture by mass of molten zinc and cadmium is shown in the diagram below.

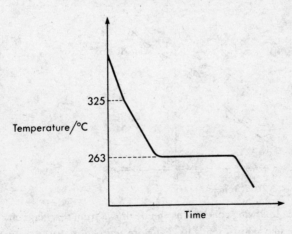

Which of the following statements is/are correct?
1. The melting point of pure zinc is 325 °C.
2. Solid begins to separate at 325 °C.
3. Pure cadmium separates at 263 °C.
4. The eutectic temperature is 263 °C.

20 The temperature–composition diagram for the phenol–water system is shown below.

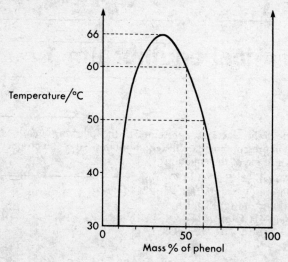

Which of the following conclusions can be drawn from this diagram?
1 Phenol and water are completely miscible above 66 °C.
2 A 50:50 phenol–water mixture is completely miscible above 60 °C.
3 At 30 °C a solution containing approximately 10% of phenol by mass in water is in equilibrium with a solution containing approximately 30% of water by mass in phenol.
4 A 60:40 phenol–water mixture is completely miscible at 40 °C.

7 Chemical equilibrium 1

1. In which one of the reactions represented by the following equations will an increase in pressure increase the equilibrium yield of the products on the right-hand side of the equation?
   A $CaCO_3(s) \rightleftharpoons CaO(s) + CO_2(g)$
   B $4HCl(g) + O_2(g) \rightleftharpoons 2H_2O(g) + 2Cl_2(g)$
   C $2HI(g) \rightleftharpoons H_2(g) + I_2(g)$
   D $3Fe(s) + 4H_2O(g) \rightleftharpoons Fe_3O_4(s) + 4H_2(g)$
   E $NH_4Cl(g) \rightleftharpoons NH_3(g) + HCl(g)$

2. Which one of the following affects the value of the equilibrium constant (K) of a reversible reaction?
   A addition of a catalyst
   B change in concentration of a reactant
   C change in concentration of a product
   D change in total pressure
   E change in temperature

3. The simplest expression for the equilibrium constant (K_c) for the reaction represented by the equation
$$2NH_3(g) + 3CuO(s) \rightleftharpoons 3Cu(s) + N_2(g) + 3H_2O(g)$$
is

   A $\dfrac{[Cu][N_2][H_2O]}{[NH_3][CuO]}$ B $\dfrac{[Cu]^3[N_2][H_2O]^3}{[NH_3]^2[CuO]^3}$

   C $\dfrac{[N_2][H_2O]}{[NH_3]}$ D $\dfrac{[N_2][H_2O]^3}{[NH_3]^2}$

   E $\dfrac{[N_2]}{[NH_3]^2}$

7 Chemical equilibrium 1

4 The degree of dissociation of dinitrogen tetroxide

$$N_2O_4(g) \rightleftharpoons 2NO_2(g)$$

at temperature T and total pressure p is α. Which one of the following is the correct expression for the equilibrium constant (K_p) at this temperature?

A $\quad \dfrac{2\alpha}{1-\alpha^2}$

B $\quad \dfrac{\alpha^2 p}{1-\alpha}$

C $\quad \dfrac{4\alpha^2 p}{1-\alpha}$

D $\quad \dfrac{4\alpha^2}{1-\alpha^2}$

E $\quad \dfrac{4\alpha^2 p}{1-\alpha^2}$

5 Which one of the following is the correct expression for the solubility product (K_{sp}) of bismuth(III) sulphide, Bi_2S_3?
  A $\quad [Bi^{3+}][S^{2-}]$
  B $\quad [Bi^{3+}]^2[S^{2-}]$
  C $\quad [Bi^{3+}]^2[S^{2-}]^3$
  D $\quad [Bi^{3+}][S^{2-}]^3$
  E $\quad [Bi^{3+}]^3[S^{2-}]^2$

6 The solubility product (K_{sp}) of silver(I) chromate(VI) at 298 K is given by

$$2Ag^+(aq) + CrO_4^{2-}(aq) \rightleftharpoons Ag_2CrO_4(s) \qquad K_{sp} = 1.0 \times 10^{-12} \text{ mol}^3 \text{ dm}^{-9}.$$

What is the concentration of silver(I) ion in a saturated solution of silver(I) chromate(VI) at this temperature?
  A $\quad 1.0 \times 10^{-12}$ mol dm^{-3}
  B $\quad 3.6 \times 10^{-8}$ mol dm^{-3}
  C $\quad 1.0 \times 10^{-6}$ mol dm^{-3}
  D $\quad 1.3 \times 10^{-4}$ mol dm^{-3}
  E $\quad 1.0 \times 10^{-4}$ mol dm^{-3}

7 Chemical equilibrium 1

SITUATION SETS

Select the response from **A–E** that correctly answers each of the items in the set. Each letter may be used ONCE, MORE THAN ONCE or NOT AT ALL.

Items 7 to 10

The main chemical reaction occurring in the contact process for the manufacture of sulphuric(VI) acid may be represented by the equation

$$2SO_2(g) + O_2(g) \underset{}{\overset{\text{exothermic}}{\rightleftharpoons}} 2SO_3(g).$$

7 Which one of the following would increase the equilibrium yield of SO_3?

- **A** Add a catalyst.
- **B** Increase the partial pressure of oxygen.
- **C** Increase the temperature.
- **D** Decrease the total pressure.
- **E** Add an 'inert' gas to the reactants.

8* If 8.00 kg of SO_2 are completely converted to SO_3 how many mol of SO_3 will be produced?

- **A** 0.0625
- **B** 0.125
- **C** 62.5
- **D** 125
- **E** 250

9* If 50 dm³ of $SO_2(g)$ react completely to give $SO_3(g)$ what is the volume of $O_2(g)$ required? (All volumes are measured at the same temperature and pressure.)

- **A** 25 dm³
- **B** 50 dm³
- **C** 100 dm³
- **D** 11.2 dm³
- **E** 22.4 dm³

10 Which one of the following correctly represents the change (if any) in the oxidation number of the sulphur atom in the forward reaction?

- **A** −II to −III
- **B** −IV to −VI
- **C** +II to +III
- **D** +IV to +VI
- **E** no change

7 Chemical equilibrium 1

Items 11 to 13

A sample of hydrogen iodide is 50% dissociated by volume at a certain temperature T:

$$2HI(g) \rightleftharpoons H_2(g) + I_2(g).$$

11 Starting with 2.0 mol of HI the number of mol of H_2 at equilibrium is
- **A** 0.125
- **B** 0.25
- **C** 0.50
- **D** 1.0
- **E** 2.0

12 The equilibrium constant (K_c) for this reaction at temperature T is equal to
- **A** 0.0625
- **B** 0.25
- **C** 0.50
- **D** 1.0
- **E** 4.0

13 If a mixture of 1.0 mol of H_2 and 1.0 mol of I_2 is allowed to reach equilibrium at temperature T the number of mol of HI present will be
- **A** 0.125
- **B** 0.25
- **C** 0.50
- **D** 1.0
- **E** 2.0

Test Paper 7 continued overleaf

7 Chemical equilibrium 1

MULTIPLE COMPLETION ITEMS

ONE or MORE THAN ONE of the four responses numbered 1–4 may be correct. Consider each of the responses carefully and decide whether or not it is correct. Mark your answer sheet as follows.

A	B	C	D	E
only 1, 2 and 3 correct	only 1 and 3 correct	only 2 and 4 correct	only 4 correct	some other response, or combination of responses, correct

14 Which of the following descriptions of the state of chemical equilibrium reached in a reversible reaction is/are correct?
1. The rate of the forward reaction has fallen to zero.
2. The reactants have been completely converted to products.
3. Only certain reactant molecules are capable of reaction, leaving the others unchanged.
4. The rates of the forward and reverse reactions are equal.

15 The reaction of bromine with water may be represented by the equation

$$Br_2(l) + H_2O(l) \rightleftharpoons Br^-(aq) + OBr^-(aq) + 2H^+(aq).$$

The addition of which of the following would move the equilibrium position of this reaction to the right?
1. potassium carbonate
2. sodium bromide
3. sodium hydroxide
4. sulphuric(VI) acid

16 When solutions of iron(III) chloride and potassium thiocyanate (KSCN) are mixed a red colour due to the complex ion $[FeSCN]^{2+}$ is produced:

$$Fe^{3+}(aq) + SCN^-(aq) \rightleftharpoons [FeSCN]^{2+}(aq).$$

The addition of which of the following would intensify the colour?
1. ammonium chloride
2. iron(II) chloride
3. potassium chloride
4. potassium thiocyanate

7 Chemical equilibrium 1

17 In the reversible reaction represented by the equation

$$3Fe(s) + 4H_2O(g) \rightleftharpoons Fe_3O_4(s) + 4H_2(g)$$

the equilibrium partial pressure of hydrogen can be increased by
1. adding a catalyst.
2. increasing the partial pressure of steam.
3. increasing the total pressure.
4. using a larger amount of iron.

18 In which of the reactions represented by the following equations is the equilibrium constant in terms of concentrations (K_c) equal to that in terms of partial pressures (K_p)?
1. $Cl_2(g) \rightleftharpoons 2Cl(g)$
2. $N_2O(g) + Mg(s) \rightleftharpoons MgO(s) + N_2(g)$
3. $2NO(g) + Cl_2(g) \rightleftharpoons 2NOCl(g)$
4. $H_2(g) + I_2(g) \rightleftharpoons 2HI(g)$

Test Paper 7 continued overleaf

7 Chemical equilibrium 1

ASSERTION–REASON ITEMS

A statement (*assertion*) is followed by a *reason*. Consider the assertion on its own and decide whether it is a true statement. Then consider the reason on its own and decide whether it is a true statement. If you decide that BOTH the assertion AND the reason are true, consider whether the reason is a correct explanation of the assertion. Mark your answer sheet as follows.

	Assertion	Reason	
A	True	True	Reason is a CORRECT EXPLANATION of the assertion
B	True	True	Reason is NOT A CORRECT EXPLANATION of the assertion
C	True	False	
D	False	True	
E	False	False	

	Assertion		*Reason*
19	When dilute hydrochloric acid is added to a sulphate(IV) (sulphite), sulphur dioxide is evolved	BECAUSE	the equilibrium position of the reaction $$SO_3^{2-}(aq) + 2H^+(aq) \rightleftharpoons SO_2(g) + H_2O(l)$$ is moved to the right when $H^+(aq)$ ions are added.
20	The equilibrium constant for the reaction represented by the equation $$CO(g) + 2H_2(g) \rightleftharpoons CH_3OH(g)$$ increases as the total pressure rises	BECAUSE	according to Le Chatelier's principle an increase in pressure moves the equilibrium position of a reversible reaction in the direction which involves a decrease in volume.

8 Chemical equilibrium 2: redox equilibria and electrochemistry

1. An unknown metal **M** displaces nickel from nickel(II) sulphate(VI) solution, but does not displace manganese from manganese(II) sulphate(VI) solution. Which one of the following represents the correct order of reducing power (most powerful first) of the three metals?

   A manganese, nickel, **M**
   B manganese, **M**, nickel
   C nickel, manganese, **M**
   D nickel, **M**, manganese
   E **M**, nickel, manganese

2. The electrolysis of aqueous sodium chloride using a mercury cathode results in the discharge at the cathode of sodium ions rather than hydrogen ions. Which one of the following is the best explanation for this observation?

   A A liquid cathode is used.
   B Sodium is above hydrogen in the reactivity series.
   C Hydrogen has a high overvoltage at a mercury cathode.
   D The standard electrode potential of sodium is more negative than that of hydrogen.
   E The discharge potential of sodium ions is greater than that of hydrogen ions at a mercury cathode.

3. Which one of the following species is oxidised when a concentrated solution of potassium hydrogensulphate(VI) is electrolysed between inert electrodes in the production of hydrogen peroxide?

   A H^+(aq)   B H_2O C K^+ D OH^- E HSO_4^-

8 Chemical equilibrium 2

4 4.0×10^{-4} mol of xenon trioxide (XeO_3) are required for the complete oxidation of 24 cm^3 of a solution which is 0.020 M with respect to manganese(II) ion to manganate(VII) (permanganate) ion (MnO_4^-). The final oxidation state of xenon in this reaction is

    A 0
    B +II
    C +III
    D +IV
    E +V

5 When 1×10^5 dm^3 of ordinary water is electrolysed under suitable conditions until only 1 dm^3 of water remains it is found that this contains 99% 'heavy water' (D_2O). Which one of the following values is the most likely for the standard electrode potential $E^\ominus$ of D^+/D_2?

    A -0.3 V
    B -0.003 V
    C 0
    D $+0.003$ V
    E $+0.3$ V

SITUATION SET

Select the response from **A–E** that correctly answers each of the items in the set. Each letter may be used ONCE, MORE THAN ONCE or NOT AT ALL.

Items 6 to 9

The standard electrode potential data ($E^\ominus$) for various copper and iron species are shown in the table below.

Electrode	$E^\ominus$/V
$Cu^{2+}(aq)/Cu(s)$	$+0.34$
$Cu^+(aq)/Cu(s)$	$+0.52$
$Cu^{2+}(aq)/Cu^+(aq)$	$+0.15$
$Fe^{3+}(aq)/Fe(s)$	-0.04
$Fe^{2+}(aq)/Fe(s)$	-0.44
$Fe^{3+}(aq)/Fe^{2+}(aq)$	$+0.77$

6 Which one of the following species is the most powerful reducing agent under standard conditions?

    A $Cu(s)$
    B $Cu^+(aq)$
    C $Fe(s)$
    D $Fe^{2+}(aq)$
    E $Fe^{3+}(aq)$

8 Chemical equilibrium 2

7 Which one of the following species would be expected to oxidise Fe^{2+}(aq) to Fe^{3+}(aq) under standard conditions?
  A   Cu(s)
  B Cu^+(aq)
  C Cu^{2+}(aq)
  D   Fe(s)
  E   none of these

8 What is the standard e.m.f. of the cell

$$Fe(s)/Fe^{2+}(aq)\|Cu^{2+}(aq)/Cu(s)?$$

  A   0.10 V
  B   0.30 V
  C   0.38 V
  D   0.78 V
  E   0.96 V

9 The electrode potential E for the system Cu^{2+}(aq, 0.1 M)/Cu(s) would be expected to be ($RT/F = 0.06$ under the experimental conditions)
  A   +0.28 V
  B   +0.31 V
  C   +0.34 V
  D   +0.37 V
  E   +0.40 V

Test Paper 8 continued overleaf

8 Chemical equilibrium 2

MULTIPLE COMPLETION ITEMS

ONE or MORE THAN ONE of the four responses numbered 1–4 may be correct. Consider each of the responses carefully and decide whether or not it is correct. Mark your answer sheet as follows.

A	B	C	D	E
only 1, 2 and 3 correct	only 1 and 3 correct	only 2 and 4 correct	only 4 correct	some other response, or combination of responses, correct

10 Which of the following equations represent(s) a redox reaction?
1. $Ag^+(aq) + Cl^-(aq) \rightarrow AgCl(s)$
2. $PCl_5(s) + H_2O(l) \rightarrow POCl_3(l) + 2H^+(aq) + 2Cl^-(aq)$
3. $Cu^{2+}(aq) + 4Cl^-(aq) \rightarrow [CuCl_4]^{2-}(aq)$
4. $[Fe(CN)_6]^{4-}(aq) + \frac{1}{2}Cl_2(g) \rightarrow [Fe(CN)_6]^{3-}(aq) + Cl^-(aq)$

11 Which of the reactions represented by the following equations can be classified as disproportionation?
1. $2C_2H_5\cdot(g) \rightarrow C_2H_6(g) + C_2H_4(g)$
2. $2H_2O_2(aq) \rightarrow 2H_2O(l) + O_2(g)$
3. $3MnO_4^{2-}(aq) + 4H^+(aq) \rightarrow 2MnO_4^-(aq) + MnO_2(s) + 2H_2O(l)$
4. $2N_2O_5(g) \rightarrow 2N_2O_4(g) + O_2(g)$

12* Which of the following are possible values for the charge on 1 mol of a metal ion?
1. 4.82×10^4 C
2. 9.65×10^4 C
3. 14.5×10^4 C
4. 19.3×10^4 C

13 When the same quantity of electricity is passed through two solutions containing a nickel salt and a silver salt respectively, 0.02 mol of nickel is deposited on one cathode and 0.04 mol of silver on the other cathode. Which of the following conclusions can be drawn *from these observations alone*?
1. Both nickel and silver ions are positively charged.
2. The nickel ion carries two units of charge.
3. The charge on the nickel ion is twice that on the silver ion.
4. The silver ion carries one unit of charge.

8 Chemical equilibrium 2

14 The electrolytic conductivity κ of a weak acid HA in 0.01 M aqueous solution is 1.6×10^{-4} S cm^{-1} and the molar conductivity at 'infinite dilution' Λ_∞ of the acid is 400 S cm^2 mol^{-1}. Which of the following conclusions can be drawn from these data?

1. The molar conductivity of the 0.01 M solution is 16 S cm^2 mol^{-1}.
2. The acid is monobasic.
3. The degree of dissociation of the acid in 0.01 M solution is 0.040.
4. The molar conductivity of the acid is higher at greater dilutions.

15 Which of the following factors influence the magnitude of the standard electrode potential of a halogen X_2?

1. The bond dissociation energy of X–X.
2. The electron affinity of X.
3. The hydration energy of the X$^-$ ion.
4. The ionisation energy of X.

16 The standard electrode potentials $E^\ominus$ of three metals in water and in liquid ammonia are shown in the table below.

Electrode	$E^\ominus$(H$_2$O)/V	$E^\ominus$(NH$_3$)/V
K$^+$/K	−2.93	−1.98
Ca^{2+}/Ca	−2.87	−1.74
Na$^+$/Na	−2.71	−1.85

Which of the following conclusions can be drawn from these data?

1. Calcium is a stronger reducing agent than sodium in water.
2. Calcium is a weaker reducing agent than sodium in liquid ammonia.
3. Potassium is the strongest reducing agent in both water and liquid ammonia.
4. Potassium is a stronger reducing agent in liquid ammonia than in water.

Test Paper 8 continued overleaf

8 Chemical equilibrium 2

ASSERTION–REASON ITEMS

A statement (*assertion*) is followed by a *reason*. Consider the assertion on its own and decide whether it is a true statement. Then consider the reason on its own and decide whether it is a true statement. If you decide that BOTH the assertion AND the reason are true, consider whether the reason is a correct explanation of the assertion. Mark your answer sheet as follows.

	Assertion	Reason	
A	True	True	Reason is a CORRECT EXPLANATION of the assertion
B	True	True	Reason is NOT A CORRECT EXPLANATION of the assertion
C	True	False	
D	False	True	
E	False	False	

	Assertion		Reason
17	The ion-electron equation $ClO^- + 2H^+ + 2e^- \rightarrow Cl^- + H_2O$ represents a reduction process	BECAUSE	when chlorate(I) (hypochlorite) ion is converted to chloride ion electrons are gained.
18	The quantity of electricity required to liberate 1 mol of copper atoms from copper(II) sulphate(VI) solution is twice that required to liberate 1 mol of silver atoms from silver(I) nitrate(V) solution	BECAUSE	the charge on a copper ion in copper(II) sulphate(VI) solution is +2 and that on a silver ion in silver(I) nitrate(V) solution is +1.
19	The platinum used in the standard hydrogen electrode is covered with a layer of finely divided platinum	BECAUSE	platinum is used in the standard hydrogen electrode to make electrical connection with the solution.
20	The electrode potential for the system $Pt, H_2(g)/H^+(aq, 0.1\ M)$ is positive	BECAUSE	a decrease in hydrogen ion concentration displaces the equilibrium $H^+(aq) + e^- \rightleftharpoons \tfrac{1}{2}H_2(g)$ to the left.

9 Chemical equilibrium 3: acid–base equilibria

1. Which one of the following 1 M aqueous solutions would you expect to have the highest pH value?
   - A   ammonia
   - B   ethanoic acid
   - C   hydrochloric acid
   - D   sodium carbonate
   - E   sodium hydroxide

2. Which one of the following indicators is most suitable for the titration of ethanoic acid with sodium hydroxide solution?
   - A   bromophenol blue (pH range 3.0–4.6)
   - B   bromothymol blue (pH range 6.0–7.6)
   - C   methyl red (pH range 4.4–6.2)
   - D   phenol red (pH range 6.8–8.4)
   - E   thymolphthalein (pH range 9.0–10.5)

3. Which one of the following salts gives an acidic aqueous solution?
   - A   ammonium ethanoate
   - B   barium nitrate(V)
   - C   copper(II) sulphate(VI)
   - D   potassium cyanide
   - E   sodium sulphate(VI)

Test Paper 9 continued overleaf

9 Chemical equilibrium 3

4 Which one of the following curves correctly shows the variation in pH during the titration of 0.10 M hydrochloric acid with 25 cm³ of 0.10 M sodium carbonate solution?

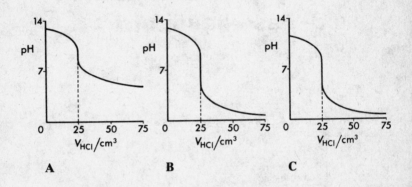

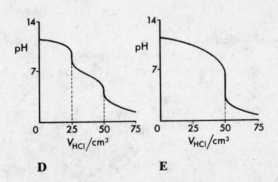

5 When 25 cm³ of 0.050 M phosphoric(v) acid (H_3PO_4) is titrated with 0.10 M sodium hydroxide solution in the presence of phenolphthalein it is found that the indicator changes colour when 25 cm³ of the alkali has been added. Which one of the following conclusions can be drawn from this observation?

    A Phosphoric(v) acid is a weak acid.
    B Phosphoric(v) acid is a diprotic (dibasic) acid.
    C The salt NaH_2PO_4 is present at the end-point.
    D The salt Na_2HPO_4 is present at the end-point.
    E The salt Na_3PO_4 is present at the end-point.

9 Chemical equilibrium 3

6 Which one of the graphs below correctly represents the change in electrical conductivity when an excess of lead(II) carbonate is gradually added to dilute nitric(V) acid?

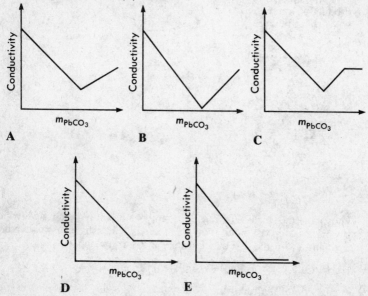

7 The ionic product for various solvents at 25 °C is given in the table below. In which one of these solvents does the neutral point occur at the lowest pH?

	Solvent	Ionic product/mol² dm⁻⁶
A	Ammonia	1×10^{-22}
B	Ethanoic acid	1×10^{-13}
C	Methanol	1×10^{-17}
D	Sulphuric(VI) acid	1×10^{-4}
E	Water	1×10^{-14}

8 The percentage dissociation of ethanoic acid (dissociation constant $K_a = 1.8 \times 10^{-5}$ mol dm⁻³) in 0.10 M solution is approximately
   A 1.8×10^{-4}
   B 1.8×10^{-2}
   C 1.3×10^{-2}
   D 0.13
   E 1.3

9 Chemical equilibrium 3

9 Which one of the following correctly expresses the relationship between the dissociation constant of a base (K_b) and the dissociation constant of its conjugate acid (K_a) at the same temperature? (K_w denotes the ionic product of water.)

A $K_b = K_a \times K_w$

B $K_b = \dfrac{1}{K_a \times K_w}$

C $K_b = \dfrac{K_w}{K_a}$

D $K_b = \dfrac{K_a}{K_w}$

E $K_b = \sqrt{(K_a \times K_w)}$

10 Phosphoric(v) acid (H_3PO_4) is a polyprotic (polybasic) acid which undergoes stepwise ionisation in aqueous solution. The dissociation constant K_1 for the first of these steps is shown below:

$$H_3PO_4(aq) \rightleftharpoons H^+(aq) + H_2PO_4^-(aq) \quad K_1 = 7.4 \times 10^{-3} \text{ mol dm}^{-3}.$$

Which one of the following is the most likely value for the dissociation constant for the second step K_2?

A 6.2×10^{-2} mol dm^{-3}
B 6.2×10^{-3} mol dm^{-3}
C 6.2×10^{-4} mol dm^{-3}
D 6.2×10^{-8} mol dm^{-3}
E 6.2×10^{-14} mol dm^{-3}

CLASSIFICATION SET

Items 11 to 14

The dissociation constants K_a of ethanoic, propanoic, chloroethanoic, iodoethanoic and trichloroethanoic acids are shown in the table below, but not necessarily in that order.

Acid	K_a/mol dm^{-3}
A	1.3×10^{-5}
B	1.7×10^{-5}
C	6.8×10^{-4}
D	1.4×10^{-3}
E	2.2×10^{-1}

Select the lettered acid **A–E** that fits the description given in each of the following items. Each letter may be used ONCE, MORE THAN ONCE or NOT AT ALL.

11 propanoic acid

12 trichloroethanoic acid

13 iodoethanoic acid

14 an acid with a pH value of approximately 1.2 in 0.10 M aqueous solution

Test Paper 9 continued overleaf

9 Chemical equilibrium 3

MULTIPLE COMPLETION ITEMS

ONE or MORE THAN ONE of the four responses numbered 1–4 may be correct. Consider each of the responses carefully and decide whether or not it is correct. Mark your answer sheet as follows.

A	B	C	D	E
only 1, 2 and 3 correct	only 1 and 3 correct	only 2 and 4 correct	only 4 correct	some other response, or combination of responses, correct

15 In which of the reactions represented by the following equations does nitric(V) acid act as a base?
1 $HNO_3 + H_2O \rightarrow H_3O^+ + NO_3^-$
2 $HNO_3 + HF \rightarrow H_2NO_3^+ + F^-$
3 $HNO_3 + CH_3COOH \rightarrow CH_3COOH_2^+ + NO_3^-$
4 $HNO_3 + 2H_2SO_4 \rightarrow NO_2^+ + 2HSO_4^- + H_3O^+$

16 Which of the following species can be classified as amphoteric, i.e. capable of showing both acidic and basic properties?
1 H_2O
2 $H_2PO_4^-$
3 $CH_2(NH_2)COOH$
4 NH_4^+

17 Which of the following would exactly neutralise 100 cm³ of 1 M sulphuric(VI) acid?
1 0.1 mol of $Ba(OH)_2$
2 0.1 mol of NaOH
3 0.1 mol of Na_2CO_3
4 0.1 mol of NH_3

18 Which of the following solutions require 25 cm³ of 0.10 M sodium hydroxide solution for complete neutralisation?
1 25 cm³ of 0.10 M chloric(VII) acid ($HClO_4$)
2 25 cm³ of 0.10 M ethanoic acid
3 25 cm³ of 0.10 M nitric(V) acid
4 25 cm³ of 0.10 M sulphuric(VI) acid

9 Chemical equilibrium 3

19 In an acid–base titration 0.10 M base is added to 25 cm³ of 0.10 M acid and the variation in pH during the titration is shown in the graph below.

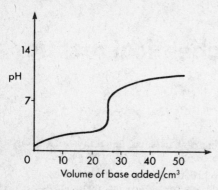

This graph could represent the titration of which of the following acid–base pairs?

1. ethanoic acid/ammonia
2. ethanoic acid/sodium hydroxide
3. hydrochloric acid/sodium hydroxide
4. nitric(V) acid/ammonia

20 A solution which is 0.10 M with respect to both ethanoic acid and sodium ethanoate acts as a buffer solution. Which of the following statements about this solution is/are correct? (pK_a for ethanoic acid = $-\lg$ (dissociation constant K_a) = 4.75.)

1. The pH of the solution is 4.75.
2. Addition of a strong acid leads to the reaction
   $$CH_3COO^-(aq) + H^+(aq) \rightarrow CH_3COOH(aq):$$
3. The solution is resistant to change in pH when dilute acid or alkali is added.
4. The pH of the buffer solution increases if the concentration of both ethanoic acid and sodium ethanoate is increased to 0.20 M.

10 Rates of chemical reactions

1. Which one of the following best explains the action of a catalyst in speeding up a chemical reaction?
   - **A** It increases the equilibrium constant for the reaction.
   - **B** It increases the kinetic energy of the reacting molecules.
   - **C** It prevents the reverse reaction from occurring.
   - **D** It decreases the activation energy for the reaction.
   - **E** It decreases the enthalpy change for the reaction, i.e. ΔH becomes more negative.

2. Which one of the following best explains the rapid increase in the rate of a chemical reaction as the temperature rises?
   - **A** The collision frequency of the molecules increases as the temperature rises.
   - **B** The molecular collisions become more violent with increasing temperature.
   - **C** A considerably higher proportion of the molecules has the necessary minimum energy to react at higher temperatures.
   - **D** The bonds in the reacting molecules are more readily broken as the temperature rises.
   - **E** The transition state is more easily converted to the products with increasing temperature.

3. Which one of the following organic chloro-compounds would be hydrolysed most rapidly under given conditions?
   - **A** $CH_3CH_2CH_2CH_2Cl$
   - **B** $CH_3CH_2CH(Cl)CH_3$
   - **C** $(CH_3)_2CHCH_2Cl$
   - **D** $(CH_3)_3CCl$
   - **E** C_6H_5Cl

10 Rates of chemical reactions

4 Which one of the graphs below correctly represents the effect of temperature on the rate of an enzyme-catalysed reaction?

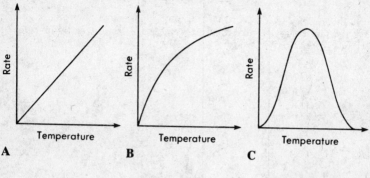

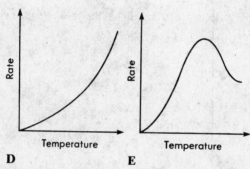

5 The following results were obtained in an investigation of the effect of temperature on the rate of a reaction by the 'clock' method.

Temperature/°C	Time of clock/s
16	400
40	50
56	12.5

The rate of this reaction is doubled by an increase in temperature of
- **A** 8 °C
- **B** 10 °C
- **C** 12 °C
- **D** 16 °C
- **E** 24 °C

10 Rates of chemical reactions

6 The activation energy for the platinum-catalysed reaction represented by the equation

$$2H_2O_2(aq) \rightarrow 2H_2O(l) + O_2(g)$$

in the absence of a catalyst is 75 kJ mol^{-1}. The most likely value for the activation energy for this reaction in the presence of a platinum catalyst is

- **A** 8.0 kJ mol^{-1}
- **B** 49 kJ mol^{-1}
- **C** 68 kJ mol^{-1}
- **D** 75 kJ mol^{-1}
- **E** 98 kJ mol^{-1}

CLASSIFICATION SET

Items 7 to 9

The energy profile diagrams for a reaction in the presence and in the absence of a catalyst are shown below.

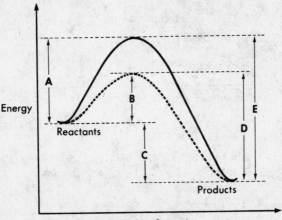

Select the appropriate enthalpy (energy) change (labelled **A–E**) for each of the following items. Each letter may be used ONCE, MORE THAN ONCE or NOT AT ALL.

7 Enthalpy change for the reaction.

8 Activation energy of the uncatalysed forward reaction.

9 Activation energy of the catalysed reverse reaction.

10 Rates of chemical reactions

SITUATION SETS

Select the response from **A–E** that correctly answers each of the items in the set. Each letter may be used ONCE, MORE THAN ONCE or NOT AT ALL.

Items 10 to 12

The acid-catalysed reaction of propanone with iodine may be represented by the equation

$$CH_3COCH_3 + I_2 \xrightarrow{H^+} CH_2ICOCH_3 + H^+ + I^-.$$

The reaction is found experimentally to be first order with respect to both propanone and hydrogen ion and zero order with respect to iodine.

10 What type of reaction is this?
  - **A** acid–base
  - **B** addition
  - **C** condensation
  - **D** elimination
  - **E** substitution

11 If the reaction were carried out under certain conditions and then repeated under the same conditions but with the concentrations of propanone, iodine and acid all doubled, what effect would this have on the initial rate of the reaction?
  - **A** 2×
  - **B** 4×
  - **C** 6×
  - **D** 8×
  - **E** none

12 Which one of the following conclusions *cannot* be drawn from these observations?
  - **A** The reaction must be a stepwise one.
  - **B** Iodine cannot be involved in the rate-determining step.
  - **C** The acid catalyst makes available a new reaction path.
  - **D** The reaction is first order overall.
  - **E** The rate equation for the reaction is

    rate = $k[CH_3COCH_3][H^+]$.

10 Rates of chemical reactions

Items 13 to 15

The decomposition of dinitrogen pentoxide in tetrachloromethane solution may be represented by the equation

$$2N_2O_5 \rightarrow 4NO_2 + O_2(g)$$

The nitrogen dioxide is soluble in tetrachloromethane. The graph below shows the decomposition curve for this reaction.

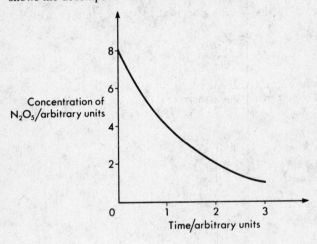

13 The order of this reaction is
    A zero
    B 1
    C 2
    D >2
    E fractional

14 Which one of the graphs opposite correctly represents the variation of the concentration of nitrogen dioxide with time?

15 The measurement of which one of the following physical quantities could *not* be used to determine the rate of this reaction?
    A volume of oxygen evolved
    B electrical conductivity of the solution
    C absorbance of the solution using a colorimeter
    D mass of reacting mixture
    E pressure of oxygen evolved

10 Rates of chemical reactions

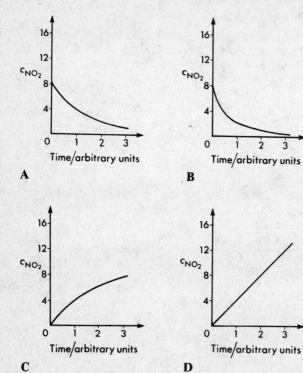

A

B

C

D

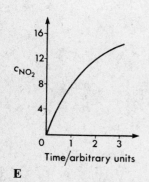

E

Test Paper 10 continued overleaf

10 Rates of chemical reactions

MULTIPLE COMPLETION ITEMS

ONE or MORE THAN ONE of the four responses numbered 1–4 may be correct. Consider each of the responses carefully and decide whether or not it is correct. Mark your answer sheet as follows.

A	B	C	D	E
only 1, 2 and 3 correct	only 1 and 3 correct	only 2 and 4 correct	only 4 correct	some other response, or combination of responses, correct

16 The action of a heterogeneous catalyst may be explained in terms of
  1 adsorbing reactant molecules on its surface.
  2 lowering the bond energies in the reactant molecules.
  3 providing an alternative reaction path.
  4 increasing the average energy of the reactant molecules.

17 Which of the following statements about the reaction of hydrogen with chlorine in the presence of light is/are correct?
  1 The rate equation for the reaction is
     rate $= k[H_2][Cl_2]$.
  2 The absorption of light causes chlorine molecules to decompose to chlorine atoms.
  3 The light lowers the activation energy for the reaction.
  4 The reaction proceeds by a radical chain mechanism.

18 The experimentally determined rate equation for the reaction represented by the equation
$$BrO_3^- + 5Br^- + 6H^+ \rightarrow 3Br_2 + 3H_2O$$
is   rate $= k[BrO_3^-][Br^-][H^+]^2$.

From this it follows that
  1 the reaction is first order with respect to bromide ion.
  2 the rate of reaction can be expressed as $-d[BrO_3^-]/dt$.
  3 doubling the concentration of BrO_3^- ion doubles the rate of reaction.
  4 doubling the concentration of each reactant multiplies the rate of reaction by eight.

10 Rates of chemical reactions

19 The experimentally determined rate equation for the oxidation of iodide ion by hydrogen peroxide in acidic solution

$$H_2O_2(aq) + 2I^-(aq) + 2H^+(aq) \rightarrow I_2(aq) + 2H_2O(l)$$

is   rate $= k_1[H_2O_2][I^-] + k_2[H^+][H_2O_2][I^-]$.

Which of the following statements about this reaction is/are correct?
1. The reaction occurs by two different paths.
2. The rate of reaction can be determined by measuring the electrical conductivity of the solution.
3. The reaction is second order with respect to iodide ion.
4. The rate of reaction is independent of pH.

20 Propan-2-ol is converted to propanone by reaction with an acidified dichromate(VI) solution:

$$3CH_3\overset{*}{C}H(OH)CH_3 + Cr_2O_7^{2-} + 8H^+ \rightarrow 3CH_3COCH_3 + 2Cr^{3+} + 7H_2O.$$

Propan-2-ol with ordinary hydrogen (^1H) in the position marked * reacts approximately six times as fast as propan-2-ol containing deuterium (^2H) in this position. Which of the following conclusions can be drawn from this information?

1. The conversion of propan-2-ol to propanone can be classified as oxidation.
2. The rate of reaction can be determined by a colorimetric method.
3. The rate-determining step must involve breaking the C—$\overset{*}{H}$ bond.
4. The reaction is first order with respect to hydrogen ion.

11 s- and p-block elements

1. Which one of the following electron configurations does *not* represent an atom of an alkali metal in the ground state?
   - A 2.8.8.1
   - B 2.8.18.1
   - C 2.8.18.8.1
   - D 2.8.18.18.8.1
   - E 2.8.18.32.18.8.1

2. In which one of the following reactions is hydrogen acting as an oxidising agent?
   - A with ethene to form ethane
   - B with iodine to form hydrogen iodide
   - C with nitrogen to form ammonia
   - D with sodium to form sodium hydride
   - E with sulphur to form hydrogen sulphide

3. Francium (atomic number 87) is an alkali metal in Group I of the periodic table. Which one of the following conclusions *cannot* be drawn from this?
   - A Francium is a solid at room temperature.
   - B The electronegativity of francium is lower than that of any of the other alkali metals.
   - C The first ionisation energy of francium is lower than that of any of the other alkali metals.
   - D Francium forms an ion with a single positive charge.
   - E Francium forms a cation with a smaller radius than that of any of the other alkali metals.

11 s- and p-block elements

4 The electron configuration of an atom of the element radium can be represented as [noble gas] $7s^2$. Which one of the following statements is *unlikely* to be correct?

    A Radium has an oxidation number of +II in all its compounds.
    B Radium decomposes water at room temperature, liberating hydrogen.
    C Radium carbonate is more stable than calcium carbonate with respect to thermal decomposition.
    D Radium hydroxide is amphoteric.
    E Radium sulphate(VI) is sparingly soluble in water.

5 In which one of the compounds represented by the following formulae does the nitrogen atom have the lowest oxidation number?

    A NH_3
    B NH_2OH
    C HNO_3
    D HNO_2
    E NO

6 When ammonia gas is dissolved in water the following reaction occurs:

$$NH_3(g) + H_2O(l) \rightleftharpoons NH_4^+(aq) + OH^-(aq).$$

In which one of the following ways is the water acting in this reaction?

    A an acid
    B a base
    C an oxidising agent
    D a reducing agent
    E an inactive solvent

7 Which one of the following statements about the hydrolysis of the trihalides of the Group V elements is correct?

    A Nitrogen trichloride is resistant to hydrolysis.
    B Nitrogen trifluoride gives ammonia, hydrogen fluoride and oxygen.
    C Phosphorus trichloride gives phosphine and hydrogen chloride.
    D Bismuth(III) chloride gives bismuth(III) chloride oxide (BiOCl) and hydrogen chloride.
    E Hydrolysis of the trichlorides occurs more rapidly as the atomic number of the element increases.

11 s- and p-block elements

8 The products of the hydrolysis of sulphur dichloride oxide (SOCl$_2$) with aqueous sodium hydroxide are

- **A** sulphur dioxide and hydrogen chloride.
- **B** sulphur(VI) oxide and hydrogen chloride.
- **C** sulphur dioxide, sodium chloride and water.
- **D** sodium sulphate(VI), sodium chloride and water.
- **E** sodium sulphate(IV) (sulphite), sodium chloride and water.

9 Fluorine reacts with water at room temperature to liberate mainly oxygen according to the equation

$$2F_2(g) + 2H_2O(l) \rightarrow 4HF(g) + O_2(g).$$

In which one of the following ways is the water acting in this reaction?

- **A** an acid
- **B** a base
- **C** an oxidising agent
- **D** a reducing agent
- **E** a solvent

CLASSIFICATION SET

Items 10 to 13

The five halogen elements (in order of increasing atomic number) are

- **A** fluorine
- **B** chlorine
- **C** bromine
- **D** iodine
- **E** astatine

Select the lettered halogen **A–E** that has the property indicated in each of the following items. Each letter may be used ONCE, MORE THAN ONCE or NOT AT ALL.

10 Liquid at standard temperature and pressure.

11 Used in the manufacture of solvents.

12 Highest first ionisation energy.

13 Has a naturally-occurring radioactive isotope.

11 s- and p-block elements

MULTIPLE COMPLETION ITEMS

ONE or MORE THAN ONE of the four responses numbered 1–4 may be correct. Consider each of the responses carefully and decide whether or not it is correct. Mark your answer sheet as follows.

A	B	C	D	E
only 1, 2 and 3 correct	only 1 and 3 correct	only 2 and 4 correct	only 4 correct	some other response, or combination of responses, correct

14 In which of the following properties does hydrogen resemble a metal?
 1 It forms a diatomic molecule H_2.
 2 Solid hydrogen has a hexagonal close-packed structure.
 3 It forms solid salt-like hydrides containing the H^- ion.
 4 It forms an ion that is discharged at the cathode in electrolysis.

15 Which of the following properties of lithium and its compounds illustrate its diagonal relationship with magnesium?
 1 Lithium forms Li_2O rather than Li_2O_2 when it is burnt in oxygen.
 2 Lithium carbonate is sparingly soluble in water.
 3 Lithium nitrate(V) decomposes on heating to give nitrogen dioxide and oxygen.
 4 Lithium salts are generally hydrated.

16 The compounds of which of the following elements do *not* give a flame colouration?
 1 barium
 2 calcium
 3 lithium
 4 magnesium

17 Cyanogen $((CN)_2)$ is sometimes described as a 'pseudohalogen' and the cyanide group (CN) as a 'pseudohalide' as they are similar in some respects to halogens and halide ions respectively. In which of the following properties does this apply?
 1 The compound of the cyanide group with hydrogen is a very weak acid.
 2 Cyanogen reacts with alkalis to give cyanide and cyanate (OCN^-) ions.
 3 Cyanide ion forms a hexacoordinate complex ion with iron(II) ion, $[Fe(CN)_6]^{4-}$.
 4 Silver(I) cyanide is sparingly soluble in water.

11 s- and p-block elements

18 Which of the following general methods for the preparation of hydrides could be used for silane (SiH₄)?
   1. direct combination of silicon with hydrogen
   2. hydrolysis of a metal silicide with water or dilute acid
   3. reaction of a metal silicide with a strong, non-volatile acid
   4. reduction of silicon tetrachloride with lithium tetrahydridoaluminate

19 Selenium is placed below sulphur in Group VI of the periodic table of the elements. Which of the following conclusions about selenium and its compounds can be drawn from this?
   1. Selenium forms a gaseous hydride H_2Se.
   2. Selenium forms an acidic oxide SeO_3.
   3. The principal oxidation states of selenium are II, IV and VI.
   4. Selenium is a gas at room temperature and pressure.

20 Group trends in the properties of the halogen elements in the order fluorine, chlorine, bromine and iodine include an increase in
   1. boiling point of the element.
   2. electronegativity.
   3. ionic radius of the halide ion.
   4. oxidising power of the element.

12 d-block elements

1 Which one of the following is *not* a characteristic property of a d-block element or its compounds?
   - A catalytic activity
   - B formation of coloured ions
   - C formation of complex ions
   - D high standard electrode potential
   - E paramagnetism

2 When an excess of aqueous ammonia is added to a solution of a zinc salt the white precipitate formed initially dissolves to give a colourless solution. Which one of the following formulae correctly represents the complex ion formed in this reaction?
   - A $[Zn(NH_3)_4]^{2+}$
   - B $[ZnO_2]^{2-}$
   - C $[Zn(OH)_4]^{2-}$
   - D $[Zn(H_2O)_6]^{2+}$
   - E $[Zn(OH)_4(H_2O)_2]^{2-}$

3 What is the oxidation number of the nickel atom in the complex salt $K_4Ni(CN)_4$?
   - A 0
   - B +II
   - C +IV
   - D +VI
   - E +VIII

12 d-block elements

4 Which one of the following formulae correctly represents the complex ion formed when a solution of potassium cyanide is added to a solution of an iron(II) salt?

- A $[Fe(H_2O)_6]^{2+}$
- B $[Fe(H_2O)_6]^{3+}$
- C $[Fe(OH)_6]^{4-}$
- D $[Fe(CN)_6]^{3-}$
- E $[Fe(CN)_6]^{4-}$

5 In which one of the following compounds is the oxidation number of the vanadium atom highest?

- A $VSO_4 \cdot 7H_2O$
- B $VO.SO_4$
- C $V_2(SO_4)_3 \cdot 3H_2O$
- D NH_4VO_3
- E $K_4V(CN)_6 \cdot 3H_2O$

6 The following reaction scheme summarises the laboratory preparation of potassium manganate(VII) (permanganate) starting from a manganese(II) salt:

$$Mn^{2+}(aq) \xrightarrow[\text{I}]{ClO_3^-} MnO_4^{2-}(aq) \xrightarrow[\text{II}]{Cl_2} MnO_4^-(aq) + MnO_2(s)$$

pale pink   green   purple   black

Which one of the following statements is *not* correct?

- A Both steps I and II involve the oxidation of manganese.
- B The oxidation number of the manganese atom in the MnO_4^{2-} ion is +VI.
- C Only one half of the original manganese can theoretically be converted to potassium manganate(VII).
- D The manganese(IV) oxide formed in step II can be removed by filtration.
- E Manganese undergoes disproportionation in step II.

7 Gold can have an oxidation number of either +I or +III in its compounds. In gold(III) compounds the bonding is mainly covalent. Which one of the following is the most likely structure for a chloride of gold with the empirical formula $AuCl_2$ in the light of this information?

- A $Au^{2+}(Cl^-)_2$
- B Cl–Au–Cl
- C $Au^+[AuCl_4]^-$
- D $(Au^+)_2[AuCl_6]^{2-}$
- E $(AuCl_2)_n$ (polymeric)

12 d-block elements

8 Only one third of the total chlorine in a compound with the empirical formula $CrCl_3 \cdot 6H_2O$ can be precipitated by silver nitrate solution at room temperature. Which one of the following is the most likely structure for the compound?

- **A** $[CrCl(H_2O)_5]^{2+}(Cl^-)_2 \cdot H_2O$
- **B** $[CrCl_2(H_2O)_4]^+ Cl^- \cdot 2H_2O$
- **C** $[CrCl_3(H_2O)_3] \cdot 3H_2O$
- **D** $[Cr(H_2O)_6]^{3+}(Cl^-)_3$
- **E** $Cr^{3+}(Cl^-)_3 \cdot 6H_2O$

9 The absorption spectrum in the visible region of a solution of a titanium(III) salt in dilute hydrochloric acid is shown in the diagram below.

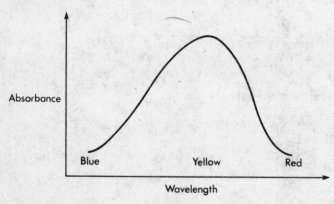

The colour of the solution is most likely to be

- **A** green
- **B** orange
- **C** purple
- **D** red
- **E** yellow

Test Paper 12 continued overleaf

12 d-block elements

10 In a colorimetric investigation of the complex ion formed by a metal ion M^{2+} and a neutral ligand L the absorbance of light by the solution of the complex ion is measured for different concentrations of M^{2+} and L, their total concentration being kept constant. When the results are plotted a curve of the form shown below is obtained.

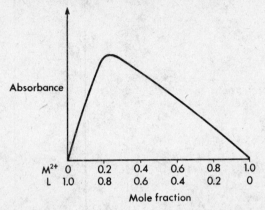

Which one of the following formulae correctly represents the complex ion?

   A $[ML_2]^+$
   B $[ML_2]^{2+}$
   C $[ML_4]^+$
   D $[ML_4]^{2+}$
   E $[ML_6]^{2+}$

MULTIPLE COMPLETION ITEMS

ONE or MORE THAN ONE of the four responses numbered 1–4 may be correct. Consider each of the responses carefully and decide whether or not it is correct. Mark your answer sheet as follows.

A	B	C	D	E
only 1, 2 and 3 correct	only 1 and 3 correct	only 2 and 4 correct	only 4 correct	some other response, or combination of responses, correct

11 In which of the following compounds is the oxidation number of the nickel atom zero?

   1 $KNiF_3$
   2 $Ni(CO)_4$
   3 $Ni(NH_3)_6Cl_2$
   4 $Ni(PF_3)_4$

12 d-block elements

12 In which of the following pairs of compounds of d-block elements is the oxidation number of the d-block element the same?
1 Co(NH$_3$)$_6$Cl$_3$ and K$_2$CoCl$_4$
2 K$_2$CrO$_4$ and K$_2$Cr$_2$O$_7$
3 K$_4$Fe(CN)$_6$ and K$_3$FeF$_6$
4 MnSO$_4$.4H$_2$O and Mn(CH$_3$COO)$_2$.7H$_2$O

13 Which of the following methods could be used to determine the number of ions present in a solution of a complex salt?
1 electrical conductivity
2 freezing point depression
3 infrared spectroscopy
4 osmotic pressure

14 In an experimental investigation of the reaction between mercury(II) and iodide ions the height of the precipitate produced is measured (after centrifuging). Different volumes of 1 M aqueous potassium iodide are added to a series of identical test tubes each containing 5.0 cm^3 of 1 M aqueous mercury(II) chloride. When the results are plotted a curve of the form shown below is obtained.

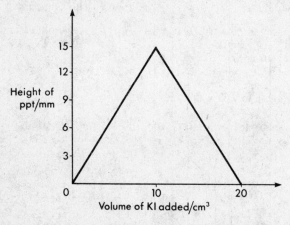

Which of the following conclusions can be drawn from these data?
1 The formula of the precipitate is HgI$_2$.
2 The height of the precipitate decreases after the addition of 10 cm^3 of KI(aq) due to complex ion formation.
3 A complex ion of formula [HgI$_4$]$^{2-}$ is formed in the reaction.
4 If the addition of KI(aq) is continued beyond 20 cm^3 the precipitate should reappear.

12 d-block elements

ASSERTION–REASON ITEMS

A statement (*assertion*) is followed by a *reason*. Consider the assertion on its own and decide whether it is a true statement. Then consider the reason on its own and decide whether it is a true statement. If you decide that BOTH the assertion AND the reason are true, consider whether the reason is a correct explanation of the assertion. Mark your answer sheet as follows.

	Assertion	Reason	
A	True	True	Reason is a CORRECT EXPLANATION of the assertion
B	True	True	Reason is NOT A CORRECT EXPLANATION of the assertion
C	True	False	
D	False	True	
E	False	False	

Assertion — *Reason*

15 The maximum oxidation number shown by a manganese atom in its compounds is +VII BECAUSE the manganese atom can use its five 3d and two 4s electrons as valence electrons.

16 Silver(I) chloride dissolves in aqueous ammonia to give a colourless solution BECAUSE silver(I) chloride forms a colourless, water-soluble complex with ammonia.

17 When concentrated hydrochloric acid is gradually added to aqueous copper(II) sulphate(VI) the colour of the solution changes from blue through green to yellow BECAUSE copper(II) chloride is formed in the reaction between copper(II) sulphate(VI) and hydrochloric acid.

18 The ethanedioate (oxalate) ion in the complex ion BECAUSE the iron atom in the $[Fe(C_2O_4)_3]^{3-}$ complex ion has an oxidation number of +VI.

[structural diagram of $[Fe(C_2O_4)_3]^{3-}$ complex ion, charge 3−]

is hexadentate

12 d-block elements

	Assertion		Reason
19	The Cu^+ ion is paramagnetic in its compounds	BECAUSE	the Cu^+ ion contains an unpaired d electron.
20	A tetrahedral complex of the type Ma_2b_2 can exist in *cis* and *trans* forms	BECAUSE	the structures $\begin{smallmatrix} & a & \\ b-&M&-a \\ & b & \end{smallmatrix}$ and $\begin{smallmatrix} & a & \\ b-&M&-b \\ & a & \end{smallmatrix}$ are different.

13 Organic chemistry 1: structure and bonding

1 Which one of the structures below represents an isomer of

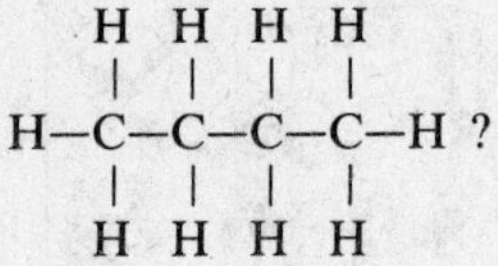

```
    H H H H
    | | | |
H—C—C—C—C—H ?
    | | | |
    H H H H
```

A
```
    H H H
    | | |
H—C—C—C—H
    | |  |
    H H  |
       H—C—H
         |
         H
```

B
```
    H H H
    | | |
H—C—C—C—H
    |  |  |
    H  |  H
     H—C—H
       |
       H
```

C
```
    H
    |
  H—C—H
           H
           |
  H—C———C—H
    |     |
    H   H—C—H
          |
          H
```

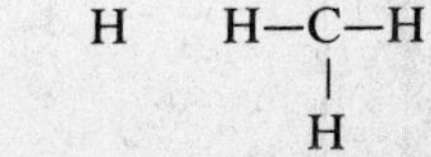

D
```
    H H
    | |
H—C—C—H
    | |
H—C—C—H
    | |
    H H
```

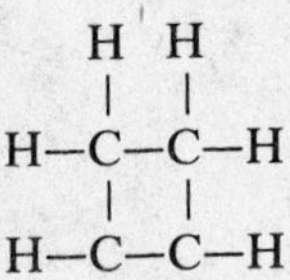

E
```
         H
         |
    H  H—C—H
    |   |
 H—C———C—H
     \ /
      C
     / \
    H   H
```

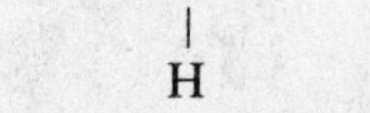

80

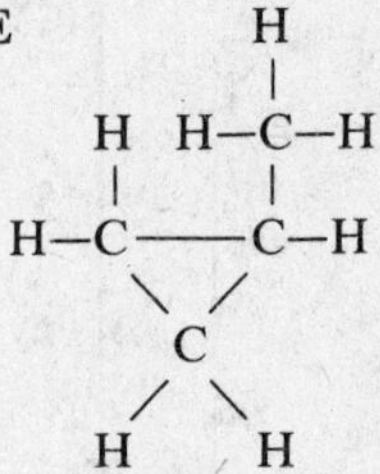

13 Organic chemistry 1

2. Which one of the diagrams below most accurately represents the variation in the boiling point of the unbranched alkanes with the number of carbon atoms (n_C) in the molecule?

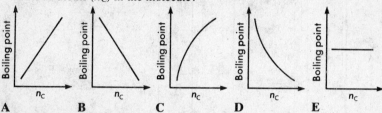

3. Which one of the following liquids is likely to have the highest viscosity at a given temperature?

   A pentane
   B pentan-1-ol
   C pentan-2-ol
   D pentane-1,5-diol
   E pentane-1,3,5-triol

4. Which one of the following organic liquids is most likely to have zero dipole moment?

   A benzene-1,2-diol
   B 1,3-dichlorobenzene
   C 1,4-dinitrobenzene
   D ethoxybenzene
   E phenylamine

5.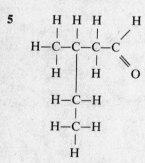

   The correct systematic name for the compound above is

   A 2-ethylbutanal.
   B 3-ethylbutanal.
   C 2-methylpentanal.
   D 3-methylpentanal.
   E hexanal.

13 Organic chemistry 1

6 Which one of the following statements about organic reactions is correct?
- **A** They generally take place very slowly.
- **B** They generally occur in a series of steps.
- **C** They involve a change in the nuclear structure of one or more carbon atoms.
- **D** They do not involve ions.
- **E** They can be classified as either addition or substitution.

7 The products of the reaction of chloromethane with 'heavy water' (D_2O) under suitable conditions are
- **A** $CH_3OH + HCl$
- **B** $CH_3OH + DCl$
- **C** $CH_3OD + HCl$
- **D** $CH_3OD + DCl$
- **E** $CD_3OH + HCl$

8 Which one of the following reagents does *not* react with ethanal?
- **A** AlH_4^-
- **B** CH_3CONH_2
- **C** CN^-
- **D** $MnO_4^- + H^+$
- **E** NH_2OH

9 Which one of the following statements about the relative acid strengths of pentanoic acid and the chloro-substituted acid
$$ClCH_2CH_2CH_2CH_2COOH$$
is correct?
- **A** Pentanoic acid is much stronger than $ClCH_2CH_2CH_2CH_2COOH$.
- **B** Pentanoic acid is slightly stronger than $ClCH_2CH_2CH_2CH_2COOH$.
- **C** Pentanoic acid and $ClCH_2CH_2CH_2CH_2COOH$ are equally strong.
- **D** $ClCH_2CH_2CH_2CH_2COOH$ is slightly stronger than pentanoic acid.
- **E** $ClCH_2CH_2CH_2CH_2COOH$ is much stronger than pentanoic acid.

10 Which one of the following products would result from the reaction of benzene with iodine chloride (ICl) in the presence of a suitable catalyst?

A [benzene with Cl substituent]

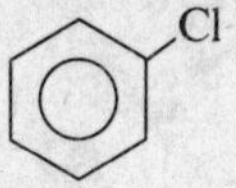

B [benzene with I substituent]

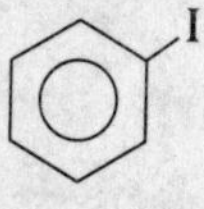

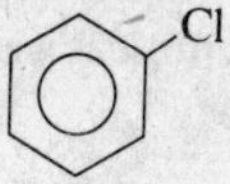

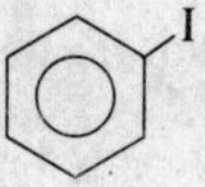

C a mixture of [chlorobenzene] and [iodobenzene]

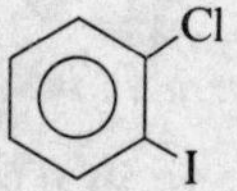

D [1,3,5-trichlorobenzene]

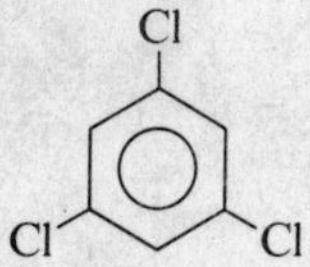

E a mixture of [1-Cl, 2-I benzene] and [1-Cl, 4-I benzene]

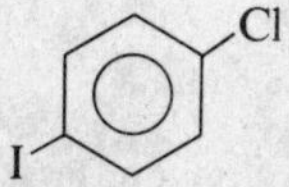

MULTIPLE COMPLETION ITEMS

ONE or MORE THAN ONE of the four responses numbered 1–4 may be correct. Consider each of the responses carefully and decide whether or not it is correct. Mark your answer sheet as follows.

A	B	C	D	E
only 1, 2 and 3 correct	only 1 and 3 correct	only 2 and 4 correct	only 4 correct	some other response, or combination of responses, correct

11 Which of the following molecular formulae *must* represent an unsaturated hydrocarbon?

1 C_6H_6
2 C_6H_{10}
3 C_6H_{12}
4 C_6H_{14}

13 Organic chemistry 1

12 Which of the following formulae represent an alkene that can exist in *cis*- and *trans*-forms?
1. $H_2C=C(CH_3)_2$
2. $CH_3CH=CHCH_3$
3. $CH_3CH=C(CH_3)_2$
4. $CH_3CH=C(CH_3)C_2H_5$

13 Which of the following formulae represent a compound that can exist as optical isomers?
1. $CH_3CH(NH_2)COOH$
2. $HOOCCH_2CH(NH_2)CH_2COOH$
3. $C_6H_5-CH(OH)CH_3$
4. $CH_3CH=CHCH_3$

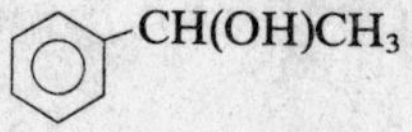

14 Which of the following diagrams correctly represent the polar carbon–chlorine bond in a chloroalkane?

1. $-\overset{\delta+}{C}-\overset{\delta-}{Cl}$
2. $-C \rightarrow Cl$
3. $-C \blacktriangleleft Cl$
4. $-C : Cl$

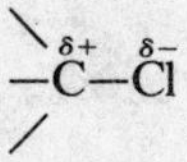

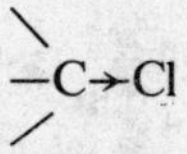

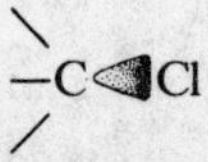

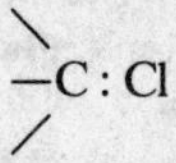

15 Which of the following formulae represent a compound that would give the alkene $CH_3CH_2CH=CHCH_3$ on dehydration?
1. $CH_3CH_2CH_2CH_2CH_2OH$
2. $CH_3CH_2CH_2CH(OH)CH_3$
3. $(CH_3)_2CHCH_2CH_2OH$
4. $CH_3CH_2CH(OH)CH_2CH_3$

16 The mechanism of the alkaline hydrolysis of 2-bromo-2-methylpropane is thought to involve the following two steps

$$\underset{\underset{Br}{|}}{\overset{\overset{CH_3}{|}}{CH_3-C-CH_3}} \xrightarrow{\text{slow}} \underset{+}{\overset{\overset{CH_3}{|}}{CH_3-C-CH_3}} + Br^-$$

$$\underset{+}{\overset{\overset{CH_3}{|}}{CH_3-C-CH_3}} \xrightarrow[\text{fast}]{OH^-} \underset{\underset{OH}{|}}{\overset{\overset{CH_3}{|}}{CH_3-C-CH_3}}$$

Which of the following observations support this mechanism?

1 The rate of hydrolysis is independent of the hydroxide ion concentration.
2 The rate of hydrolysis is proportional to the concentration of $(CH_3)_3CBr$.
3 A tertiary carbonium ion such as $(CH_3)_3C^+$ is relatively stable.
4 The $(CH_3)_3CBr$ molecule has an appreciable dipole moment.

Test Paper 13 continued overleaf

13 Organic chemistry 1

ASSERTION–REASON ITEMS

A statement (*assertion*) is followed by a *reason*. Consider the assertion on its own and decide whether it is a true statement. Then consider the reason on its own and decide whether it is a true statement. If you decide that BOTH the assertion AND the reason are true, consider whether the reason is a correct explanation of the assertion. Mark your answer sheet as follows.

	Assertion	Reason	
A	True	True	Reason is a CORRECT EXPLANATION of the assertion
B	True	True	Reason is NOT A CORRECT EXPLANATION of the assertion
C	True	False	
D	False	True	
E	False	False	

	Assertion		*Reason*
17	The reaction between methane and chlorine is usually carried out in the presence of light	BECAUSE	light catalyses the reaction between methane and chlorine.
18*	The boiling point of ethanol is lower than that of ethanethiol (C_2H_5SH)	BECAUSE	the relative molecular mass of ethanol is lower than that of ethanethiol.
19	Phenol is a stronger acid than ethanol	BECAUSE	the ethanol molecule is stabilised by delocalisation which reduces the proton-donating ability of the —OH group.
20	Benzene undergoes addition rather than substitution reactions	BECAUSE	the benzene molecule is unsaturated.

14 Organic chemistry 2: chemistry of functional groups

1. A compound has the following properties:
   (i) it reacts with phosphorus pentachloride with the evolution of hydrogen chloride,
   (ii) it reacts with sodium hydroxide solution to form an ionic solid,
   (iii) it undergoes an addition reaction with hydrogen in the presence of a suitable catalyst.

   Which one of the following formulae correctly represents this compound?

   A $CH_3CH_2CH_2OH$
   B CH_3COOH
   C $C_6H_5CH_2OH$
   D $CH_2{=}CHCH_2OH$
   E $CH_2{=}CHCOOH$

2. Which one of the following equations correctly represents the reaction of benzene with ethanoyl chloride in the presence of a suitable catalyst?

   A $C_6H_6 + CH_3COCl \rightarrow C_6H_5Cl + CH_3CHO$
   B $C_6H_6 + CH_3COCl \rightarrow C_6H_5CH_3 + CO + HCl$
   C $C_6H_6 + CH_3COCl \rightarrow C_6H_5Cl + CO + CH_4$
   D $C_6H_6 + CH_3COCl \rightarrow C_6H_5CHO + CH_3Cl$
   E $C_6H_6 + CH_3COCl \rightarrow C_6H_5COCH_3 + HCl$

3. Ethanal (CH_3CHO) is manufactured by the catalytic oxidation of ethene by air. Which one of the following formulae correctly represents the compound which would be formed from propene under similar conditions?

   A $CH_3CH\underset{O}{\overset{}{-\!\!\!\diagdown\;\diagup\!\!\!-}}CH_2$
   B $CH_3CH_2CH_2CHO$
   C CH_3COCH_3

   D $CH_3CH_2CH_2OH$
   E $CH_3CH(OH)CH_2OH$

14 Organic chemistry 2

CLASSIFICATION SETS

Items 4 to 7

Five types of organic reaction are given below.

    A   Electrophilic addition
    B   Electrophilic substitution
    C   Nucleophilic addition
    D   Nucleophilic substitution
    E   Radical reaction

Select the lettered reaction type **A–E** that correctly describes each of the following reactions. Each letter may be used ONCE, MORE THAN ONCE or NOT AT ALL.

4   $CH_3CH_2CH_2CH_2Br \xrightarrow{OH^-} CH_3CH_2CH_2CH_2OH$

5   $CH_3CH_2CHO \xrightarrow{CN^-/H^+} CH_3CH_2CH(OH)CN$

6   $CH_3CH_2CH{=}CH_2 \xrightarrow{Br_2} CH_3CH_2CHBrCH_2Br$

Items 8 to 11

The structural formulae of five aromatic compounds are given below.

   CH₂OH   OH   COCH₃   CONH₂   Cl

(phenyl rings)

   A   B   C   D   E

Select the lettered compound **A–E** that could be characterised by means of the reagents given in the following items. Each letter may be used ONCE, MORE THAN ONCE or NOT AT ALL.

8   Iron(III) chloride solution.

9   2,4-Dinitrophenylhydrazine.

10   Phosphorus pentachloride.

11   Nitric(III) acid (nitrous acid).

14 Organic chemistry 2

MULTIPLE COMPLETION ITEMS

ONE or MORE THAN ONE of the four responses numbered 1–4 may be correct. Consider each of the responses carefully and decide whether or not it is correct. Mark your answer sheet as follows.

A	B	C	D	E
only 1, 2 and 3 correct	only 1 and 3 correct	only 2 and 4 correct	only 4 correct	some other response, or combination of responses, correct

12 Which of the following properties is/are characteristic of the homologous series of alkanes?
  1 general formula C_nH_{2n+2}
  2 unbranched chain structure
  3 similar physical properties
  4 similar chemical properties

13 Which of the compounds represented by the following formulae would be formed in the reaction of ethene with aqueous bromine in the presence of sodium chloride?
  1 CH_2BrCH_2Br
  2 CH_2BrCH_2Cl
  3 CH_2BrCH_2OH
  4 CH_2ClCH_2Cl

14 An organic compound with the molecular formula $C_5H_{12}O$ on prolonged oxidation gives a compound with the molecular formula $C_5H_{10}O_2$. Which of the following are possible structural formulae for the original organic compound?
  1 $CH_3CH_2CH_2CH_2CH_2OH$
  2 $CH_3CH_2CH(CH_3)CH_2OH$
  3 $(CH_3)_3CCH_2OH$
  4 $CH_3CH_2CH(OH)CH_2CH_3$

15 Which of the compounds represented by the following formulae give a yellow precipitate on warming with iodine and sodium hydroxide solution?
  1 CH_3CH_2OH
  2 $CH_3CH(OH)CH_3$
  3 $CH_3COCH_2CH_3$
  4 $CH_3CH_2COCH_2CH_3$

14 Organic chemistry 2

16 A colourless compound gives a yellow precipitate with 2,4-dinitrophenylhydrazine reagent, but does not give a silver mirror on warming with ammoniacal silver(I) nitrate(V) solution (Tollens' reagent). Which of the following are possible structural formulae for the compound?

1. $CH_3CH(OH)CH_3$
2. CH_3CH_2CHO
3. CH_3CONH_2
4. CH_3COCH_3

ASSERTION–REASON ITEMS

A statement (*assertion*) is followed by a *reason*. Consider the assertion on its own and decide whether it is a true statement. Then consider the reason on its own and decide whether it is a true statement. If you decide that BOTH the assertion AND the reason are true, consider whether the reason is a correct explanation of the assertion. Mark your answer sheet as follows.

	Assertion	Reason	
A	True	True	Reason is a CORRECT EXPLANATION of the assertion
B	True	True	Reason is NOT A CORRECT EXPLANATION of the assertion
C	True	False	
D	False	True	
E	False	False	

	Assertion		Reason
17	Halogenoalkanes react with sodium cyanide in ethanolic solution to give alkyl cyanides	BECAUSE	cyanide ion has a negative charge which attacks the electron-deficient carbon atom in a halogenoalkane.
18	In the laboratory preparation of ethanal by the oxidation of ethanol, the product is distilled as it is formed	BECAUSE	ethanal can undergo further oxidation to ethanoic acid.
19	Ethanoic acid gives a yellow precipitate with 2,4-dinitrophenylhydrazine reagent	BECAUSE	the ethanoic acid molecule contains a carbonyl group ($>C=O$).
20	The diazotisation of phenylamine (aniline) with sodium nitrate(III) (nitrite) and hydrochloric acid is usually carried out at 5–10 °C	BECAUSE	the diazonium salt produced from phenylamine at 5–10 °C decomposes to reform phenylamine at higher temperatures.

15 Organic chemistry 3: large molecules

1 Which one of the following substances is a carbohydrate?
   A cellulose
   B DNA
   C insulin
   D penicillin
   E saccharin

2 The structural formula $CH_3(CH_2)_{16}COOCH_2$
   $CH_3(CH_2)_{16}COOCH$
   $CH_3(CH_2)_{16}COOCH_2$

   represents a molecule of which one of the following types of substance?
   A a carbohydrate
   B a fat
   C a polymer
   D a protein
   E a vitamin

3 Which one of the following is *not* a polymer?
   A cellulose
   B perspex
   C a protein
   D a silicone
   E a synthetic detergent

15 Organic chemistry 3

4 The simplified structural formula

```
         base              base
          |                 |
    .....sugar—phosphate—sugar—phosphate.....
```

represents which one of the following?

- **A** an enzyme
- **B** a fat
- **C** a nucleic acid
- **D** a protein
- **E** a vitamin

5 When pieces of human hair are heated strongly with soda–lime the smell of ammonia can be detected. Which one of the following conclusions can be drawn from this observation?

- **A** Ammonia is present in hair.
- **B** An ammonium salt is present in hair.
- **C** Hair contains amino acids.
- **D** Hair contains nitrogen.
- **E** None of these conclusions is valid.

6 Which one of the following types of reaction occurs when a peptide link is broken?

- **A** condensation
- **B** elimination
- **C** hydration
- **D** hydrolysis
- **E** substitution

7 Which one of the following correctly represents the structure of the monomer from which the polymer

$$-CF_2.CF_2.CF_2.CF_2.CF_2.CF_2-$$

is obtained?

A
$$\begin{array}{c} F \\ | \\ F-C-F \\ | \\ F \end{array}$$

B
$$\begin{array}{c} F \quad F \\ | \quad | \\ C=C \\ | \quad | \\ F \quad F \end{array}$$

C
$$\begin{array}{c} F \quad F \quad F \\ | \quad | \quad | \\ F-C-C=C \\ | \qquad | \\ F \qquad F \end{array}$$

D
$$\begin{array}{c} F \quad F \quad F \quad F \\ | \quad | \quad | \quad | \\ F-C-C-C=C \\ | \quad | \quad | \\ F \quad F \quad F \end{array}$$

E
$$\begin{array}{c} F \quad F \quad F \quad F \\ | \quad | \quad | \quad | \\ F-C-C=C-C-F \\ | \qquad \quad | \\ F \qquad \quad F \end{array}$$

15 Organic chemistry 3

8 Which one of the following simple molecules is eliminated in the formation of nylon-66 from hexanedioyl dichloride ClCO(CH$_2$)$_4$COCl and hexane-1,6-diamine H$_2$N(CH$_2$)$_6$NH$_2$?

  A ammonia
  B hydrogen chloride
  C hydrogen cyanide
  D water
  E none of these

9 Which one of the following substances contains C=C bonds in its structure?

  A nylon-66
  B polytetrafluoroethene
  C rubber
  D starch
  E terylene

10 Which one of the following substances is a polyester?

  A hexane-1,6-diamine
  B nylon-66
  C polystyrene
  D polytetrafluoroethene
  E terylene

11 The main product in the ozonolysis of natural rubber is a compound with the structure CH$_3$COCH$_2$CH$_2$CHO. Which one of the following is the most likely structure of natural rubber?

  A $\left[\text{CH}-\text{CH}_2-\text{CH}_2-\text{CH}_2\right]_n$ with CH$_3$ branch
  B $\left[\text{CH}-\text{CH}_2-\text{CH}_2-\text{CH}\right]_n$ with CH$_3$ and OH branches
  C $\left[\text{CH}_2-\text{CH}=\text{CH}-\text{CH}\right]_n$ with CH$_3$ branch
  D $\left[\text{CH}_2-\text{C}=\text{C}-\text{CH}_2\right]_n$ with OH and CH$_3$ branches
  E $\left[\text{CH}_2-\text{CH}=\text{C}-\text{CH}_2\right]_n$ with CH$_3$ branch

15 Organic chemistry 3

12 In a paper chromatographic investigation of the four sugars sucrose, maltose, lactose and raffinose the chromatogram shown in the diagram below was obtained.

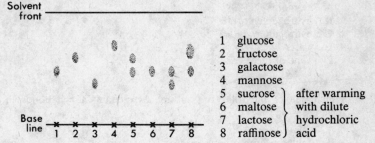

1 glucose
2 fructose
3 galactose
4 mannose
5 sucrose ⎫
6 maltose ⎬ after warming with dilute hydrochloric acid
7 lactose ⎪
8 raffinose ⎭

Which one of the following conclusions can be drawn from these observations?

A Maltose is unaffected by dilute hydrochloric acid.
B Sucrose is hydrolysed to give a mixture of glucose and mannose.
C Lactose is hydrolysed to give a mixture of glucose and galactose.
D Raffinose is hydrolysed to give a mixture of glucose and fructose.
E None of the four sugars investigated gives fructose on hydrolysis.

MULTIPLE COMPLETION ITEMS

ONE or MORE THAN ONE of the four responses numbered 1–4 may be correct. Consider each of the responses carefully and decide whether or not it is correct. Mark your answer sheet as follows.

A	B	C	D	E
only 1, 2 and 3 correct	only 1 and 3 correct	only 2 and 4 correct	only 4 correct	some other response, or combination of responses, correct

13 Which of the following substances contain nitrogen?

1 diesel oils
2 DNA
3 fats
4 proteins

14 Which of the following substances can be classified as synthetic fibres?

1 acrilan
2 cotton
3 orlon
4 silk

15 Organic chemistry 3

15 Which of the following statements about cracking crude oil is/are correct?

1. The process is often carried out at relatively low temperatures in the presence of a catalyst.
2. Some of the products are unsaturated.
3. Dehydrogenation occurs during the process.
4. Branched chain alkanes are converted to unbranched chain alkanes which are preferred in petrol.

16 Which of the following substances would react to give a polymer?

1. HO(CH$_2$)$_6$OH and ClCO(CH$_2$)$_4$COCl
2. H$_2$N(CH$_2$)$_6$NH$_2$ and H$_3$C.COO(CH$_2$)$_4$COOCH$_3$
3. H$_3$C.COO(CH$_2$)$_4$CH=CH$_2$ alone

4.
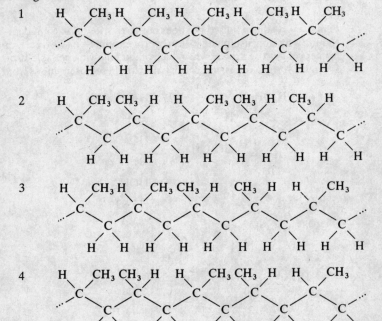

17 Which of the following polymer structures represent(s) a syndiotactic arrangement?

15 Organic chemistry 3

18 In the production of polychloroethene it is found that the use of a small amount of di(benzoyl)peroxide $C_6H_5CO.O.O.COC_6H_5$ may cause the polymerisation of a large number of molecules of chloroethene $CH_2=CHCl$. Which of the following statements about this process are likely to be correct?

1 The structure of polychloroethene may be represented as

$$\left[\begin{array}{cc} H & H \\ | & | \\ -C-C- \\ | & | \\ H & Cl \end{array} \right]_n$$

2 The reaction proceeds by a radical mechanism.
3 The di(benzoyl)peroxide acts as a chain initiator.
4 The di(benzoyl)peroxide decomposes to give benzoate ions $C_6H_5COO^-$.

19 Glucose is readily oxidised but does not give an addition compound with sodium hydrogensulphate(IV) (hydrogensulphite); several optical isomers of glucose are known. Which of the following possible structural formulae for glucose is/are consistent with these observations?

1, 2, 3, 4 — structural formulae of glucose (furanose form, open-chain ketose, pyranose form, and open-chain aldose respectively).

15 Organic chemistry 3

20 Part of the aminoacid chain in beef insulin is shown below.

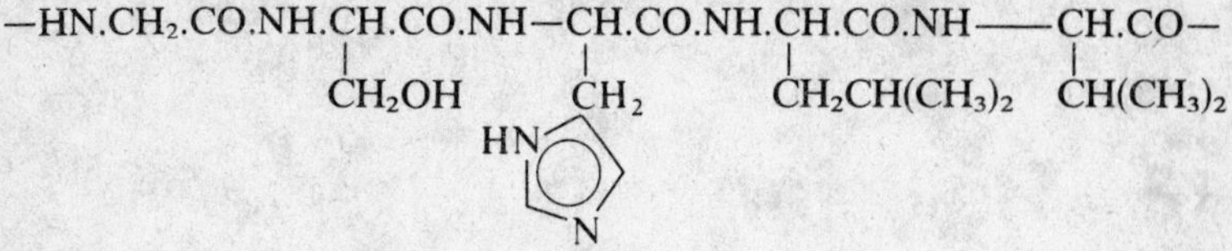

$-HN.CH_2.CO.NH.CH.CO.NH-CH.CO.NH.CH.CO.NH\text{———}CH.CO-$
  | | | |
  CH_2OH CH_2 $CH_2CH(CH_3)_2$ $CH(CH_3)_2$

(with an imidazole ring: HN—CH=N attached to the CH_2)

Which of the following aminoacids would be obtained by the complete hydrolysis of beef insulin?

1 $H_2NCHCOOH$
    |
    $CH(CH_3)_2$

2 $H_2NCHCOOH$
    |
    $CH(CH_3)CH_2CH_3$

3 $H_2NCHCOOH$
    |
    $CH_2CH(CH_3)_2$

4 $H_2NCHCOOH$
    |
    $CH(OH)CH_3$

16 Revision test

1* Which one of the following contains the largest number of atoms?
- A 3.0 g of nitrogen
- B 8.0 g of sulphur
- C 10 g of chlorine
- D 16 g of zinc
- E 50 g of lead

2* A gas contains approximately 25% nitrogen and 75% oxygen by mass. The simplest formula for the gas is
- A N_2O
- B NO
- C NO_2
- D N_2O_3
- E N_2O_5

3 100 cm³ of a solution of iron(III) sulphate(VI) in water contains 2.81 g of $Fe_2(SO_4)_3.9H_2O$. What is the concentration of this solution with respect to iron(III) ions? ($Fe_2(SO_4)_3 = 400$, $9H_2O = 162$.)
- A 0.0167 M
- B 0.025 M
- C 0.050 M
- D 0.10 M
- E 0.15 M

4* The product obtained by the loss of an α-particle from an atom of $^{238}_{92}U$ can be represented by which one of the following symbols?
- A $^{238}_{93}Np$
- B $^{238}_{91}Pa$
- C $^{242}_{94}Pu$
- D $^{234}_{90}Th$
- E $^{234}_{92}U$

5 Hydroxylammonium ion NH_3OH^+ reduces iron(III) ion to iron(II) ion in sulphuric(VI) acid solution. If 2 mol of Fe^{3+} ion is reduced by 1 mol of NH_3OH^+ ion which one of the following nitrogen-containing species is formed in the reaction?
   A ammonium ion NH_4^+
   B dinitrogen oxide N_2O
   C nitrate(III) (nitrite) ion NO_2^-
   D nitrogen N_2
   E nitrogen dioxide NO_2

6 When carbon, hydrogen, nitrogen and oxygen are allowed to remain in contact for a long time no urea can be detected:

$$C(s) + 2H_2(g) + N_2(g) + \tfrac{1}{2}O_2(g) \rightarrow CO(NH_2)_2(s) \quad \Delta H_f^\ominus = -333 \text{ kJ mol}^{-1}$$
$$\Delta G_f^\ominus = -205 \text{ kJ mol}^{-1}$$

Which one of the following is the best explanation for this observation?
   A Biochemical reactions cannot be carried out in the laboratory.
   B The enthalpy change of formation is numerically greater than the Gibbs free energy change of formation for urea.
   C The equilibrium constant for the reaction is very low.
   D The activation energy for the reaction is high.
   E The entropy change $\Delta S^\ominus$ for the reaction is positive.

7 Use the following bond energy (ΔH_b) data

$\Delta H_b / \text{kJ mol}^{-1}$ at 25 °C   C—C 348   C=C 612   C—H 412
                                                H—H 436

to calculate the enthalpy change of hydrogenation of propene (ΔH_h)

$$CH_3CH=CH_2(g) + H_2(g) \rightarrow CH_3CH_2CH_3(g)$$

at 25 °C. ΔH_h is equal to
   A -560 kJ mol^{-1}
   B -388 kJ mol^{-1}
   C -124 kJ mol^{-1}
   D $+124 \text{ kJ mol}^{-1}$
   E $+560 \text{ kJ mol}^{-1}$

Revision Test Paper continued overleaf

16 Revision test

8 The molar conductivity Λ of four dilute (1.00×10^{-3} M) aqueous solutions is given in the table below.

Compound	Λ/S cm^2 mol^{-1}
NaCl	120
BaCl$_2$	260
CeCl$_3$	400
PtCl$_4 \cdot n$NH$_3$	230

Which one of the following structural formulae for the complex platinum ammine is consistent with these data?

A [Pt(NH$_3$)$_2$Cl$_4$]
B [Pt(NH$_3$)$_3$Cl$_3$]$^+$ Cl$^-$
C [Pt(NH$_3$)$_4$Cl$_2$]$^{2+}$ 2Cl$^-$
D [Pt(NH$_3$)$_5$Cl]$^{3+}$ 3Cl$^-$
E [Pt(NH$_3$)$_6$]$^{4+}$ 4Cl$^-$

9 In which one of the following pairs are the substances isomers?

A carbon-12 and carbon-14
B diamond and graphite
C ethene and polyethene
D hexane and cyclohexane
E propanal and propanone

10* 1.4 g of a pure alkene gives 3.8 g of the corresponding dichloroalkane on reaction with chlorine under suitable conditions. Which one of the following is the correct molecular formula for the alkene?

A C_2H_4
B C_3H_6
C C_4H_8
D C_6H_{12}
E C_8H_{16}

11 The hydrolysis of ethyl ethanoate $CH_3COOC_2H_5$ is carried out in the presence of sodium hydroxide enriched with oxygen-18 (^{18}O). Which one of the following formulations of the products under these conditions is correct? ($\overset{*}{O}$ represents enrichment with oxygen-18.)

A $C_2H_5OH + CH_3C-O^- Na^+$
$\|$
O

B $C_2H_5\overset{*}{O}H + CH_3C-O^- Na^+$
$\|$
O

C $C_2H_5OH + CH_3C-\overset{*}{O}{}^- Na^+$
$\|$
O

D $C_2H_5\overset{*}{O}H + CH_3C-\overset{*}{O}{}^- Na^+$
$\|$
O

E $C_2H_5OH + CH_3C-O^- Na^+$
$\|$
$\overset{*}{O}$

12 Which one of the following properties would *not* be shown by the amino-acid with the structural formula

$HO-\langle\bigcirc\rangle-CH_2CH(NH_2)COOH$?

A alkaline in aqueous solution
B exists as two enantiomorphs
C forms a salt with hydrochloric acid
D forms a salt with sodium hydroxide
E gives a violet colour with iron(III) chloride solution

Revision Test Paper continued overleaf

16 Revision test

SITUATION SET

Select the response from **A–E** that correctly answers each of the items in the set. Each letter may be used ONCE, MORE THAN ONCE or NOT AT ALL.

Items 13 to 16

The diagram below shows the standard electrode potential $E^{\ominus}$ for various half-cell reactions involving an element X.

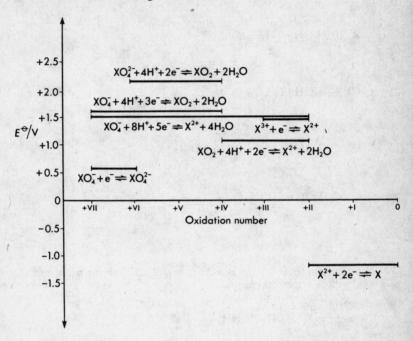

13 Which one of the following is the most likely identity of element X?
   A   chromium
   B   copper
   C   iron
   D   manganese
   E   vanadium

14 Which one of the following species is the most powerful oxidising agent?
   A X^{2+}
   B X^{3+}
   C XO_2
   D XO_4^-
   E XO_4^{2-}

16 Revision test

15 Which one of the following species is the most powerful reducing agent?

   A X
   B X^{2+}
   C X^{3+}
   D XO_2
   E XO_4^-

16 Which one of the following species would you expect to undergo disproportionation?

   A X
   B X^{2+}
   C XO_2
   D XO_4^{2-}
   E XO_4^-

CLASSIFICATION SETS

Items 17 to 20

Five types of energy change are shown below.

   A Bond dissociation energy
   B Electron affinity
   C Enthalpy change of atomisation
   D Ionisation energy
   E Lattice energy

Select the lettered energy change **A–E** that correctly describes each of the following processes. Each letter may be used ONCE, MORE THAN ONCE or NOT AT ALL.

17 $\frac{1}{2}Br_2(l) \rightarrow Br(g)$.

18 $O_2(g) \rightarrow O_2^+(g)$.

19 $NH_4^+ Cl^-(s) \rightarrow NH_4^+(g) + Cl^-(g)$.

20 A process for which the energy change cannot be measured directly.

16 Revision test

Items 21 to 24

Select the lettered graph **A–E** below that correctly represents the situation in each of the following items. Each letter may be used ONCE, MORE THAN ONCE or NOT AT ALL.

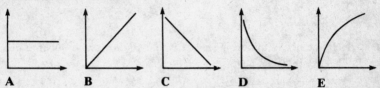

21 The product of pressure and volume for a fixed amount of an ideal gas at constant temperature (y-axis) against pressure (x-axis).

22 The square root of the frequency of corresponding lines in the X-ray spectra of different elements (y-axis) against the atomic number of the element (x-axis).

23 The molar conductivity of a strong electrolyte (y-axis) against dilution (x-axis).

24 The standard Gibbs free energy change for the reaction
$2C(s) + O_2(g) \rightarrow 2CO(g)$ (y-axis) against temperature (x-axis).

Items 25 to 28

Five bottles labelled I–V each contain a different organic compound as shown in the table below.

Bottle	Melting point/°C	Boiling point/°C
I	−129	58
II	−95	68
III	−80	151
IV	−47	158
V	43	250

Select the lettered structural formula **A–E** that correctly represents the compound described in each of the following items. Each letter may be used ONCE, MORE THAN ONCE or NOT AT ALL.

A $CH_3CH_2CH_2CH_2CH_2CH_3$
B $CH_3CH(CH_3)CH(CH_3)CH_3$
C $CH_3CH_2CH_2CH_2CH_2CH_2OH$
D $CH_3CH_2CH_2CH_2CH_2CH_2SH$
E $HOCH_2CH_2CH_2CH_2CH_2CH_2OH$

16 Revision test

25 The compound present in the bottle labelled I.

26 The compound present in the bottle labelled III.

27 The compound present in the bottle labelled V.

28 The compound which is the most soluble in water.

MULTIPLE COMPLETION ITEMS

ONE or MORE THAN ONE of the four responses numbered 1–4 may be correct. Consider each of the responses carefully and decide whether or not it is correct. Mark your answer sheet as follows.

A	B	C	D	E
only 1, 2 and 3 correct	only 1 and 3 correct	only 2 and 4 correct	only 4 correct	some other response, or combination of responses, correct

29 Which of the following pairs of substances are allotropes?
1. diamond and graphite
2. carbon-12 and carbon-14
3. oxygen (O_2) and ozone (O_3)
4. tin(II) ion (Sn^{2+}) and tin(IV) ion (Sn^{4+})

30* Which of the following represent the Avogadro constant?
1. the number of atoms in 2 g of hydrogen gas
2. the number of NH_3 molecules in 22.4 dm^3 of ammonia gas at s.t.p.
3. the number of ions in 1 dm^3 of 1 M potassium iodide solution
4. the number of electrons required to discharge 1 mol of silver(I) ions

31 Which of the following processes may be undergone by the nuclei of atoms with a neutron/proton ratio that is too low for stability?
1. beta emission
2. electron capture
3. gamma emission
4. positron emission

32 Which of the following pairs of liquids would be expected to show ideal behaviour?
1. butylamine and ethanoic acid
2. 1,1-dichloroethane and propanone
3. ethanol and cyclohexane
4. methylbenzene and 1,2-dimethylbenzene

16 Revision test

33 Which of the following statements about the reaction of a dilute solution of sodium thiosulphate(VI) with dilute hydrochloric acid

$$S_2O_3^{2-}(aq) + 2H^+(aq) \rightarrow S(s) + SO_2(g) + H_2O(l)$$

is/are correct?

1 Sulphur is precipitated in a colloidal form.
2 The thiosulphate(VI) ion undergoes disproportionation.
3 The rate of reaction can be studied by means of observations made on the precipitate.
4 The reaction can be used to distinguish between sulphate(IV) (sulphite) and thiosulphate(VI) ions.

34 The thermal decomposition of trichloroethanoate ion produces dichlorocarbene

$$CCl_3COO^- \rightarrow CCl_2 + Cl^- + CO_2.$$

Which of the following statements about dichlorocarbene (CCl_2) is/are correct?

1 Its production from trichloroethanoate ion involves an elimination reaction.
2 The carbon atom in CCl_2 is electron-deficient.
3 CCl_2 would be expected to be a non-linear species.
4 The carbon atom in CCl_2 is positively charged.

35* Which of the following properties would you expect germanium tetrachloride ($GeCl_4$) to possess?

1 conducts electricity in the liquid state
2 liquid at room temperature
3 precipitates silver(I) chloride when the liquid is shaken with solid silver(I) nitrate
4 readily hydrolysed in water

36 Which of the following reagents add on to an alkene?

1 ammonia
2 hydrogen bromide
3 sodium hydrogensulphate(IV) (hydrogensulphite)
4 sulphuric acid

37 Which of the following compounds may be formed in the reaction of ethanol with concentrated sulphuric(VI) acid under different conditions?

1 ethene
2 ethoxyethane ('ether')
3 ethyl hydrogensulphate(VI)
4 propanone

16 Revision test

38 For which of the following purposes is ethane-1,2-diol ('glycol') used?
  1 as an antifreeze in car radiators
  2 as a de-icing fluid for aeroplane wings
  3 in the manufacture of terylene
  4 in the manufacture of explosives

39 Which of the following procedures would distinguish between ethanal and propanone?
  1 Fehling's test
  2 'iodoform' (tri-iodomethane) test
  3 melting point of the 2,4-dinitrophenylhydrazine derivative
  4 'silver mirror' test (Tollens' reagent)

40 The structural formula of cyclohexanone is usually written as ⌬=O. This formula indicates that cyclohexanone
  1 is an aromatic compound.
  2 is a ketone.
  3 is unsaturated.
  4 has a molecular formula $C_6H_{10}O$.

ASSERTION–REASON ITEMS

A statement (*assertion*) is followed by a *reason*. Consider the assertion on its own and decide whether it is a true statement. Then consider the reason on its own and decide whether it is a true statement. If you decide that BOTH the assertion AND the reason are true, consider whether the reason is a correct explanation of the assertion. Mark your answer sheet as follows.

	Assertion	Reason	
A	True	True	Reason is a CORRECT EXPLANATION of the assertion
B	True	True	Reason is NOT A CORRECT EXPLANATION of the assertion
C	True	False	
D	False	True	
E	False	False	

	Assertion		Reason
41	The ion $^4He^{2+}$ would be deflected less in a mass spectrometer than the ion $^2H^+$	BECAUSE	the mass of the $^4He^{2+}$ ion is greater than that of the $^2H^+$ ion.
42	The enthalpy change of combustion of an element is equal to the enthalpy change of formation of its oxide	BECAUSE	according to Hess's law the total enthalpy change for a reaction is independent of the path taken.

16 Revision test

	Assertion		Reason
43	The weaker an acid the lower its pK_a value	BECAUSE	pK_a is defined as $\lg K_a$.
44	The rate of hydrolysis of sucrose $$C_{12}H_{22}O_{11} + H_2O \rightarrow \underset{\text{glucose}}{C_6H_{12}O_6} + \underset{\text{fructose}}{C_6H_{12}O_6}$$ can be followed by a polarimetric method	BECAUSE	as the hydrolysis of sucrose proceeds the angle of rotation of the plane of polarised light gradually changes.
45	There is only a small decrease in the size of a d-block metal atom as the atomic number rises	BECAUSE	the additional electrons in the atoms of d-block metals are entering an inner d shell.
46	An aqueous solution of potassium hexacyanoferrate(III) gives a precipitate of iron(III) hydroxide when aqueous sodium hydroxide is added	BECAUSE	iron(III) ions are present in a relatively high concentration in an aqueous solution of potassium hexacyanoferrate(III).
47	Iron(III) bromide is used as a catalyst in the bromination of aromatic compounds	BECAUSE	iron can show variable oxidation numbers in its compounds.
48	Lead may be considered to be a d-block (transition) element	BECAUSE	lead may have an oxidation number of either +II or +IV in its compounds.
49	Benzene can be nitrated using a mixture of concentrated nitric(V) and sulphuric(VI) acids	BECAUSE	a mixture of concentrated nitric(V) and sulphuric(VI) acids contains the electrophilic NO_2^+ ion.
50	The type of polymerisation involved in the manufacture of terylene is known as *condensation* polymerisation	BECAUSE	two different monomer molecules react together to give terylene, with the elimination of a small molecule.

Keys and facility indices

Item number	Paper 1 Key	Paper 1 Facility index	Paper 2 Key	Paper 2 Facility index	Paper 3 Key	Paper 3 Facility index
1	E	0.92	D	0.50	E	0.43
2	D	0.82	B	0.67	C	0.38
3	D	0.64	B	0.38	B	0.46
4	D	0.79	D	0.39	A	0.78
5	A	0.89	C	0.72	D	0.53
6	D	0.79	D	0.68	C	0.65
7	D	0.75	E	0.72	E	0.94
8	B	0.86	A	0.62	D	0.78
9	B	0.93	E	0.61	C	0.90
10	E	0.42	E	0.63	B	0.71
11	D	0.50	E	0.41	D	0.49
12	C	0.67	C	0.72	C	0.74
13	E	0.45	D	0.78	B	0.85
14	A	0.36	A	0.61	E	0.88
15	A	0.82	C	0.61	A	0.76
16	C	0.76	A	0.30	D	0.43
17	C	0.11	D	0.43	C	0.58
18	B	0.22	E(1,2,4)	0.47	E(1,2,3,4)	0.33
19	A	0.54	B	0.54	A	0.57
20	A	0.71	C	0.26	B	0.68

Mean score: 12.9   Mean score: 11.1   Mean score: 12.9

Item number	Paper 4 Key	Paper 4 Facility index	Paper 5 Key	Paper 5 Facility index	Paper 6 Key	Paper 6 Facility index
1	B	0.13	A	0.68	C	0.22
2	B	0.54	C	0.66	A	0.54
3	D	0.60	B	0.59	C	0.22
4	D	0.83	D	0.61	C	0.11
5	C	0.60	C	0.42	B	0.26
6	E	0.88	D	0.42	C	0.54
7	C	0.17	C	0.42	E	0.83
8	C	0.98	C	0.25	D	0.24
9	B	0.46	C	0.92	C	0.91
10	A	0.23	C	0.66	A	0.59
11	D	0.90	E	0.71	B	0.43
12	B	0.52	E	0.33	B	0.35
13	C	0.56	A	0.27	A	0.63
14	A	0.38	B	0.54	B	0.50
15	B	0.37	C	0.27	A	0.76
16	A	0.44	E(1,3,4)	0.59	D	0.76
17	A	0.79	B	0.85	A	0.22
18	D	0.21	A	0.86	B	0.74
19	D	0.67	C	0.37	C	0.50
20	A	0.52	A	0.10	A	0.65
	Mean score: 10.8		Mean score: 10.5		Mean score: 10.0	

	Paper 7		Paper 8		Paper 9	
Item number	Key	Facility index	Key	Facility index	Key	Facility index
1	B	0.89	B	1.00	E	0.92
2	E	0.71	C	0.41	E	0.31
3	D	0.61	E	0.33	C	0.38
4	E	0.25	A	0.53	D	0.19
5	C	0.89	B	0.37	D	0.44
6	D	0.38	C	0.47	D	0.44
7	B	0.75	E	0.65	D	0.79
8	D	0.80	D	0.76	E	0.35
9	A	0.93	B	0.20	C	0.67
10	D	0.84	D	0.51	D	0.48
11	C	0.79	A	0.27	A	0.52
12	B	0.64	C	0.82	E	0.40
13	D	0.59	B	0.90	C	0.50
14	D	0.86	E(1,2,3,4)	0.20	E	0.58
15	B	0.45	A	0.31	C	0.62
16	D	0.57	A	0.67	A	0.56
17	E(2)	0.34	A	0.76	B	0.50
18	C	0.48	A	0.76	A	0.46
19	A	0.66	B	0.29	D	0.62
20	D	0.46	D	0.55	A	0.31

Mean score: 12.9   Mean score: 10.8   Mean score: 10.0

Item number	Paper 10 Key	Paper 10 Facility index	Paper 11 Key	Paper 11 Facility index	Paper 12 Key	Paper 12 Facility index
1	D	0.98	B	0.90	D	0.87
2	C	0.75	D	0.68	A	0.48
3	D	0.43	E	0.89	A	0.90
4	C	0.56	D	0.37	E	0.77
5	A	0.90	A	0.56	D	0.85
6	B	0.66	A	0.77	C	0.40
7	C	0.98	D	0.11	C	0.60
8	A	0.98	E	0.21	B	0.75
9	D	0.92	D	0.68	C	0.67
10	E	0.72	C	0.79	D	0.81
11	B	0.69	B	0.81	C	0.62
12	D	0.57	A	0.79	C	0.69
13	B	0.61	E	0.69	E(1,2,4)	0.31
14	E	0.72	C	0.55	A	0.81
15	B	0.62	E(1,2,3,4)	0.31	A	0.96
16	B	0.77	D	0.55	A	0.50
17	C	0.51	C	0.26	C	0.33
18	A	0.66	E(2,3,4)	0.27	E	0.29
19	E(1,2)	0.52	A	0.77	E	0.31
20	A	0.49	B	0.81	E	0.54

Mean score: 14.0   Mean score: 11.8   Mean score: 12.5

	Paper 13		Paper 14		Paper 15	
Item number	Key	Facility index	Key	Facility index	Key	Facility index
1	B	0.93	E	0.71	A	0.74
2	C	0.66	E	0.69	B	0.84
3	E	0.86	C	0.60	E	0.50
4	C	0.78	D	0.83	C	0.84
5	D	0.49	C	0.56	D	0.39
6	B	0.46	A	0.52	D	0.66
7	D	0.91	B	0.63	B	0.82
8	B	0.59	B	0.42	B	0.61
9	D	0.75	C	0.67	C	0.71
10	B	0.17	A	0.60	E	0.71
11	E(1,2)	0.44	D*	(0.25)	E	0.61
12	C	0.66	E(1,4)	0.79	C	0.87
13	B	0.64	A	0.69	C	0.95
14	A	0.59	A	0.69	B	0.92
15	C	0.63	A	0.19	A	0.16
16	A	0.22	D	0.60	A	0.29
17	C	0.32	C	0.15	D	0.37
18	D	0.63	A	0.71	A	0.52
19	C	0.54	D	0.48	B	0.47
20	D	0.64	C	0.31	B	0.26

Mean score: 11.9   Mean score: 11.1   Mean score: 12.2

*This item was substantially modified as a result of pretesting.

Revision test 16

Item number	Key	Facility index	Item number	Key	Facility index
1	C	0.84	26	D	0.71
2	E	0.90	27	E	0.81
3	D	0.48	28	E	0.77
4	D	0.94	29	B	0.68
5	B	0.42	30	C	0.48
6	D	0.65	31	C	0.38
7	C	0.68	32	D	0.65
8	C	0.61	33	E(1,2,3,4)	0.38
9	E	0.52	34	A	0.16
10	B	0.81	35	C	0.61
11	C	0.45	36	C	0.65
12	A	0.58	37	A	0.48
13	D	0.84	38	A	0.61
14	E	0.65	39	E(1,3,4)	0.55
15	A	0.42	40	E(2,3,4)	0.38
16	D	0.48	41	D	0.55
17	C	0.87	42	D	0.23
18	D	0.94	43	E	0.58
19	E	0.74	44	A	0.55
20	E	0.32	45	A	0.65
21	A	0.81	46	E	0.19
22	B	0.58	47	B	0.65
23	E	0.26	48	D	0.71
24	C	0.42	49	A	0.90
25	B	0.74	50	A	0.42

Mean score: 29.7